Lecture Notes in Electrical Engineering

Volume 310

About this Series

"Lecture Notes in Electrical Engineering (LNEE)" is a book series which reports the latest research and developments in Electrical Engineering, namely:

- Communication, Networks, and Information Theory
- Computer Engineering
- Signal, Image, Speech and Information Processing
- Circuits and Systems
- Bioengineering

LNEE publishes authored monographs and contributed volumes which present cutting edge research information as well as new perspectives on classical fields, while maintaining Springer's high standards of academic excellence. Also considered for publication are lecture materials, proceedings, and other related materials of exceptionally high quality and interest. The subject matter should be original and timely, reporting the latest research and developments in all areas of electrical engineering.

The audience for the books in LNEE consists of advanced level students, researchers, and industry professionals working at the forefront of their fields. Much like Springer's other Lecture Notes series, LNEE will be distributed through Springer's print and electronic publishing channels.

More information about this series at http://www.springer.com/series/7818

Kuinam J. Kim · Naruemon Wattanapongsakorn
Editors

Mobile and Wireless Technology 2015

Springer

Editors
Kuinam J. Kim
Korea Industry Security Forum
Kyoung-gi
Korea
Republic of South Korea

Naruemon Wattanapongsakorn
King Mongkut's University of Technology Thonburi
Computer Engineering Department
Thung Khru
Bangkok
Thailand

ISSN 1876-1100 ISSN 1876-1119 (electronic)
Lecture Notes in Electrical Engineering
ISBN 978-3-662-47668-0 ISBN 978-3-662-47669-7 (eBook)
DOI 10.1007/978-3-662-47669-7

Library of Congress Control Number: 2015942634

Springer Heidelberg New York Dordrecht London

Printed on acid-free paper

Springer-Verlag GmbH Berlin Heidelberg is part of Springer Science+Business Media
(www.springer.com)

Preface

This LNEE volume contains the papers presented at the International Conference on Mobile and Wireless Technology (ICMWT2015) which was held in Bangkok, Thailand, during June 23–25, 2015.

ICMWT2015 received over 100 paper submissions from various countries. After a rigorous peer-review process, 25 full-length articles were accepted for oral presentation at the conference. This corresponds to an acceptance rate was very low and is intended for maintaining the high standards of the conference proceedings.

ICMWT2015 will provide an excellent international conference for sharing knowledge and results in Mobile and Wireless Technology. The aim of the Conference is to provide a platform to the researchers and practitioners from both academia as well as industry to meet the share cutting-edge development in the field.

The primary goal of the conference is to exchange, share and distribute the latest research and theories from our international community. The conference will be held every year to make it an ideal platform for people to share views and experiences in Mobile and Wireless Technology related fields.

On behalf of the Organizing Committee, we would like to thank Springer for publishing the proceedings of ICMWT2015. We also would like to express our gratitude to the 'Program Committee and Reviewers' for providing extra help in the review process. The quality of a refereed volume depends mainly on the expertise and dedication of the reviewers. We are indebted to the Program Committee members for their guidance and coordination in organizing the review process. and to the authors for contributing their research results to the conference.

Our sincere thanks to the Institute of Creative Advanced Technology, Engineering and Science for designing the conference web page and also spending count-less days in preparing the final program in time for printing. We would also like to thank our organization committee for their hard work in sorting our manuscripts from our authors.

We look forward to seeing all of you next year at ICMWT2016 in Korea.

Kuinam J. Kim
Kyonggi University, Korea
Naruemon Wattanapongsakorn
King Mongkut's University of Technology Thonburi, Thailand

Contents

VCloud: A Security Framework for VANET

Wiseborn Manfe Danquah and D. Turgay Altilar

Istanbul Technical University
Computer and Informatics Faculty
Department of Computer Engineering,
Istanbul –Turkey
{wdanquah,altilar}@itu.edu.tr

Abstract. Vehicular Ad-hoc Network (VANET) is an integral component of Intelligent Transport Systems (ITS) that has enjoyed a lot of attention from the research community and the automotive industry. This is mainly due to the opportunities and challenges it presents. Vehicular Ad-hoc Network being a class of Mobile Ad-hoc Networks (MANET) has all the security concerns existing in traditional MANET as well as new security and privacy concerns introduced by the unique vehicular communication environment. This paper provides a survey of the possible attacks in vehicular environment, as well as security and privacy concerns in VANET. It also provides an insight into the development of a comprehensive cloud framework to provide a more robust and secured communication among vehicular nodes and road side units. The proposed infrastructure; a Metropolitan Based Public Interconnected Vehicular Cloud (MIVC) seeks to provide a more reliable and secured vehicular communication network.

Keywords: Mobile Ad-hoc Networks, Vehicular Ad-hoc Network, Cloud, ITS, Road Side Units (RSU), Metropolitan Interconnected Vehicular Cloud (MIVC).

1 Introduction

Vehicular Ad-hoc Network is the communication among vehicles and other infrastructure along the road such as traffic lights and bill boards collectively called Road Side Units (RSU). In the VANET environment, the vehicular nodes and RSUs are equipped with communication gadgets, and on-board sensors that enable the communication.

Vehicular Ad-hoc Network is considered to be a subgroup of Mobile Ad-hoc Networks (MANET) and it is an important component of Intelligent Transport Systems (ITS) [1], which ensures that the main objective of ITS; the use of advanced and intelligent technologies to make transportation systems safer, secured and efficient, is attained.

The main objective of VANET is to deploy safety applications that reduce the number of accidents and deaths on the road. About 1.3 million people die through road accidents globally each year with an average of 3,287 deaths per day, according to the Association for Safe International Road Travel [2]. It is also predicted that by

K.J. Kim and N. Wattanapongsakorn (eds.), *Mobile and Wireless Technology 2015*,
Lecture Notes in Electrical Engineering 310, DOI: 10.1007/978-3-662-47669-7_1

2030, road accidents will be the fifth leading cause of death in the world if no remedial action is taken [2].

Vehicular Ad-hoc Network has gained a lot of attention from the research community since the late 90's after the allocation of frequency bands for different dedicated short range communications use such as vehicle-to-vehicle, vehicle-to-infrastructure communications, traffic light control and automatic toll collection. The United States Federal Communications Commission (FCC) allocated 75MHz of spectrum at 5.9GHz (from 5.850 to 5.925 GHz) of the DSRC [3] whiles their European counterpart; the European Telecommunications Standard Institute (ETSI) also allotted 30MHz of spectrum in the 5.9GHz band [4]. Despite the fact that the DSRC systems in Europe, Japan and U.S. are not currently compatible [5], the introduction of the DSRC standards has resulted in an increasing number of research with the aim of developing safety applications to make road transportation more secured and efficient.

Vehicular Ad-hoc Network applications such as driver assistance systems, road safety applications, infotainment and comfort applications do not only ensure efficient and safe road transportation but also provide a total deviation from expensive and slow traditional traffic management techniques which entails the use of cameras and sensors along roads to monitor, capture and transmit details of vehicular movement to a central point for analysis and discussion.

Even though there exit a lot of challenges in VANET such as the lack of uniform standards across regions, expensive test beds for testing algorithms and protocols, quality of service and lack of constant network access, security and privacy remain the most important challenges in VANET [6][7], security challenges in vehicular communications are exciting and unexploited topic.[8].

The goals of securing VANETs are the same as that of securing any other network; the main objective is to provide authentication, confidentiality, integrity, availability and non-repudiation [9]. Different techniques and standards have been proposed to ensure that these objectives of VANET security are achieved. In this paper a cloud computing based approach is adopted to ensure that communication in VANET environments are secured.

Cloud computing is defined by the National Institute of Standards and Technology (NIST) of the United States Department of Commerce as a model for enabling ubiquitous, convenient, on-demand network access to a shared pool of configurable computing resources that can be rapidly provisioned and released with minimal management effort or service provider interaction[10]. Deploying cloud services in VANET environment can lead to the development of new applications and services to overcome some of the challenges that exist in VANETs.

2 Security and Privacy Issues in VANET

The potential of VANET has been acknowledged with the increasing number of research seeking to address security concerns before the full deployment of VANET.

Papadimitratos et al [11] outlined the basic security requirements for VANET to include message non-repudiation, access control, message confidentiality, and privacy protection. Raya et al. [6], identified main vulnerabilities of VANET to include jamming;

where a malicious vehicular node deliberately generates transmissions that impede communications among nodes within the transmission range, Forgery; involves attackers flooding a network coverage area with false and hazardous information which may then be taken and acted upon by all receiving vehicles and Impersonation; which could exist in the network as nodes fabricate and change messages and replies in the network to make them appear as nodes they are not.

Leinmuller et al [12] also considered the sending of false node location information as one of the factors that affects the overall performance of the network. In their work, they argued that such information may lead to the formation of wrong routing links. Rawat et al [13] reviewed the different security attacks in VANET and their possible solutions and grouped VANET attacks into three main categories namely: Threat to availability, threat to Authentication and threat of confidentiality and miscellaneous attacks. Maria et al [9] also undertook a similar review of attacks in VANET and classified them as passive, insider and active attacks. Table 1 provides a summary of the various classes of attacks, examples and their vulnerabilities as well as proposed solutions.

Table 1. Classification of common attacks in VANET [9]

Attack type	Examples	Vulnerabilities	Possible solution
Passive attack	Brute force	Loss of node's identity	Secured Authentication of nodes
	Traffic analysis	Study of nodes activities by an attacker	Use of VIPER protocol
Active attack	Sybil	Multiple instance of attacker	Comparison of identity and data by neighboring nodes
	Denial of Service	Channel jamming	OBU use multiple transceivers, switch channel
	Sinkhole	Control of traffic by an attacker	Use of aggregated signature algorithm
	Position attack	Position of nodes altered or cannot be known, bogus attack	Autonomous position verification
Insider	Misbehaving attack	dropping packets and injection of faulty data by faulty nodes	Using CA and a protocol for eviction of attackers
	Illusion attack	Creation of false traffic situation by malicious nodes.	Plausibility validation network model to authenticate messages

The 1609.2 IEEE standard for Wireless Access in Vehicular Environment (WAVE) [14], has a novel way of ensuring that communication in VANET environments are more secured. Vehicular nodes are authenticated by a Certificate Authority (CA). Malicious nodes or nodes that are compromised get their key certificate revoked and a security Certificate Renovation List (CRL) is distributed to inform all legitimate nodes of this revocation thus preventing attacks from this node.

The successful distribution of the CRL to all nodes without being distorted is very significant in securing VANET communications. The distribution of the CRL is to be done solely by RSUs along the road [14]. The RSU is responsible for tracking misbehaving vehicles for certificate revocation by broadcasting CRL . The use of just RSUs in distributing CRL requires that large numbers of RSUs are deployed. In rural areas and along highways where there are not many RSUs, distribution of CRL would not be fast and efficient and malicious nodes that have not been issued a Validity

Certificate may still be able to perform attacks such as flooding the network and provide false location information.

Laberteaux et al [7] proposed a better way of resolving this problem by using vehicular nodes in the distribution of CRL. An epidemic distribution is adopted to ensure that areas where the number of RSUs do not exist or are limited would still have a faster distribution of CRL. They showed through simulations that using vehicular nodes in the distribution of CRL significantly outperforms methods that employ only RSUs as distribution points. Although this approach significantly reduces the channel load by the using incremental update of the CRL, it does not totally eliminate heavy channel load resulting from frequent broadcasting of CRL by vehicular nodes.

The use of security management scheme to replace active warning broadcasting of CRL was proposed by Samara [15]. In the proposal all vehicles will be provided with Local Revocation List (LRL) containing information about all adversaries on the current road. When nodes discover adversary nodes they move the ID of the node to the top of the LRL list for fast searching. If the Invalid Certificate (IC) of the discovered node is not in the LRL, the vehicles send details to the RSU for authentication. The RSU authenticates the adversary IC and broadcast it to all nodes. The LRL stays local for the current road and has a smaller size as compared to the CRL, thus reducing the channel overload.

Physical attack is also one of the major security concerns that have not been thoroughly studied. Malicious nodes can easily locate their targeted nodes and attack them physically by crushing into them or cause them to crush into other nodes. This may lead to multiple collisions of vehicular nodes and loss of lives. RSU's that may serves as gateway nodes to access point can also be intentionally destroyed by attackers to prevent network access within a given area.

3 An Overview of Cloud Computing

Currently, cloud computing is enjoying a lot of hype due to its numerous advantages. Companies prefer to rent from service providers the needed infrastructure and software to run their applications rather than buying hardware resources, software applications and infrastructure that may not be fully utilized or come with extra cost. Major cloud providers such as Microsoft, Amazon and IBM have experience a phenomenal increase in the number of clients in the past three years. Forecast from International Data Corporation (IDC) projects that worldwide public IT cloud services to grow from $16 billion in 2009 to $55.5 billion in 2014 (five times the rate of traditional IT services) [16]. Other areas of IT are also applying cloud technologies to make their operations more efficient and effective. The telecommunication industry for example is taking advantage of cloud technologies to ensure efficient and better service to their clients. The Telecommunications Corporation, AT&T collaborated with IBM and VMware in 2011 and introduced an enhanced mobile application that seeks to increase third-party development to deliver cloud capabilities to enhance Business Corporation and reduce IT cost [17]. The Turkish Telecommunication Company, Turkcell Superonline, have also introduced SMART cloud, a personal storage cloud system that enables users to upload all types of document on their computers. This enables clients to view photos, videos or other documents on their smartphones. It is considered as the fastest uploading service in Turkey [18].

Cloud computing employs a service oriented marketing approach where each application or resource is provided in the form of service that can be accessed by clients when it is needed. These services can be classified as Platform as a service (PaaS); a type of cloud service that provides clients with a set of tools and software that enable users to develop applications without installing development environments on their standalone computers., Software as a service (SaaS); a cloud service where clients are able to access software hosted on a cloud. SaaS includes database management software, office and messaging software and business and enterprise-oriented applications and Infrastructure as a service (IaaS): this is also termed as virtual machines and it involves providing physical resources such as servers and storage units to clients as a service. Other researchers have identified other types of services such as Desktop as a service (DaaS), Security as a service (SaaS), Networking as a service (Naas) and Collaboration as a service (CaaS)

Vehicles in regions where there are no road side units and rural areas where nodes are sparse may experience intermittent disruption in network access; however, NaaS may offer internet access to vehicles upon request at a cost. This will ensure that there is constant network access which is a requirement for most VANET applications. Collaboration as a service (CaaS) entails providing a cloud service that consists of tools that enable organizations to collaborate and share information faster and more seamlessly. In VANET collaboration as a service enabled by VCloud will ensure that autonomous cars also referred to as robotic cars [19] which are cars that are self-driving will become a reality.

4 Related Works: Cloud Computing in VANET

In this section a brief overview of previously proposed works in cloud computing in VANET is presented. Hussain et al [20] in their paper provided taxonomy of cloud in VANET. They defined and explained three classes of VANET cloud as vehicular cloud, vehicles using cloud and a hybrid that is the combination of the two techniques. They also outlined some security challenges in VANET cloud. They however did not provide possible remedy for the identified security problems. In the paper an assertion was made that despite the envisaged paradigm shift in developments in cloud computing and its applications in VANET, there is no architectural framework of cloud computing in VANET. It is in this regard that a general framework for VANET cloud to solve security problems is proposed in this paper.

Mamun et al [21], proposed a model for Ad-hoc cloud network architecture. Their proposed architecture is to enable the creation of cloud on the run model. Nodes form clusters with cluster head responsible for the communication between clusters and the formed cloud. The proposed model is more efficient in traffic control and management however the authors did not provide details of the criteria for cluster formation and cluster head selection.

Using grid technology Anda et al [22] proposed a distributed approach to solve vehicular traffic-flow control problems autonomously. The proposed vehicular grid framework basically allows the results of computation of traffic metrics by a node to be shared with its neighboring nodes. Through simulation they examined lane merging algorithms using different scenarios. It was observed that the Automatic Cruise Control implemented with the VGrid architecture improved throughput and latency of traffic merging scenarios. The results of their studies provided the motivation that, by having a central management point, resources and results of computation of one vehicular node may be shared not just between neighboring nodes but all other nodes in city-wide location that may need such resources and results of computation, which forms one of the basis for the proposed novel MIVC infrastructure in this paper.

Abuelela et al [23] in their paper taking VANET to the cloud envisaged a vehicular cloud computing system where underutilized vehicular resources such as storage, computing power and internet connectivity is shared among drivers or rented over the internet to customers. They noted Storage as a Service, (SaaS) as one of the advantages of introducing cloud computing concepts in VANETs.

Whaiduzzaman et al [24] have provided a detailed survey on vehicular cloud computing. Examples and explanations of different vehicular cloud computing services, applications of vehicular cloud, and open issues and research challenges are also outlined in the paper.

Due to the importance of security and privacy issues in VANET Yan et al [8] in their paper security challenges in vehicular cloud identified and analyzed a number of security challenges and potential privacy threats in vehicular clouds. They provided an architectural model to overcome some of the security and privacy challenges in vehicular cloud. Despite their model being able to provide a directional security scheme to overcome some challenges in VANET it does not provide a remedy to all security issues in vehicular cloud. They also did not provide a unique architectural framework for overcoming specific security challenges and recommended that future works should focus on developing specific applications that will provide feasible security and privacy solutions in vehicular clouds. The need for such a framework has led to the proposal of the Metropolitan Interconnected Vehicular Cloud (MIVC) system.

5 Contributions and Advantages of Proposed Metropolitan Interconnected Vehicular Cloud (MIVC)

The paper makes the following key contributions:

- The MIVC provides a robust infrastructure for monitoring, regulating and controlling city wide vehicular traffic.
- Proposes a novel interconnection of vehicular cloud architecture to make available cloud services in city-wide VANET environment
- Cloud services that enhance security and corporation such as Authentication as a service (AaaS) introduced can help to enforce security in Vehicular Networks.
- Other traditional cloud services such as Storage as a Service, (SaaS) and Network as a Service (NaaS) can also be provided by the proposed MIVC.

6 Metropolitan Interconnected Vehicular Cloud (MIVC)

Previous research and proposals for incorporating cloud computing into VANET environment have envisaged that single and independent public clouds would be formed and deployed to provide cloud services that are applicable in VANET environment.

Drivers of vehicles (nodes) decide to access a cloud service or not depending on the needs of the driver, type of cloud service provided, cost and the levels of security and privacy provided. Figure1. Shows a pictorial model of vehicular cloud. The clouds are formed on sections of the road where vehicles donate computing resources. Neighboring nodes can then access the cloud either directly or through a gateway node.

Fig. 1. Image showing existing proposed vehicle cloud system.

For a more accessible cloud system, a novel vehicular cloud infrastructure Metropolitan Interconnected Vehicular cloud. (MIVC). is proposed. The proposed MIVC is based on the principle that, the full potential of cloud systems in VANET environment would be more effectively utilized if there exist a city wide or metropolitan wide vehicular Cloud (VCloud) framework that would ensure that vehicular communication is more secured, uninterrupted and easily accessible to users. This metropolitan wide VCloud could provide both user specific services as well as generic services (beneficial to metropolitan /city road traffic regulators) that would ensure safer and more secured communication among vehicular nodes.

The MIVC, unlike the various proposed VCloud techniques would consist of interconnected clouds to provide a city wide cloud service that all nodes within a metropolitan area must in one way or the other (either partial or fully) be connected to. The MIVC is envisaged to be an interconnection of cloud facilities from different service providers to form one big cloud which would be administratively managed from a central point. However individual cloud service providers would be responsible for the technical operation of their cloud. The different cloud service provides could offer specific service as contribution towards the MIVC. For instance Google could provide network as a service (NaaS) in a section of the city, whiles amazon provides storage as a service (SaaS).

The management of the MIVC at a central point would be handled by the municipal or city authority for the purposes of controlling traffic which would ensure that traffic regulations are obeyed by drivers as well as ensure that cloud service providers abide by agreed upon security codes and high level of quality of service. The municipal or city authority would also be responsible for providing data concerning registered vehicles, security vehicles as well as provide authorization for cloud applications that would require conditions such as lane restrictions road diversion and road closure to be deployed.

Figure 2 shows the flow of data among the various service providers or services, RSUs and vehicular nodes (onboard equipment) is shown in the figure below.

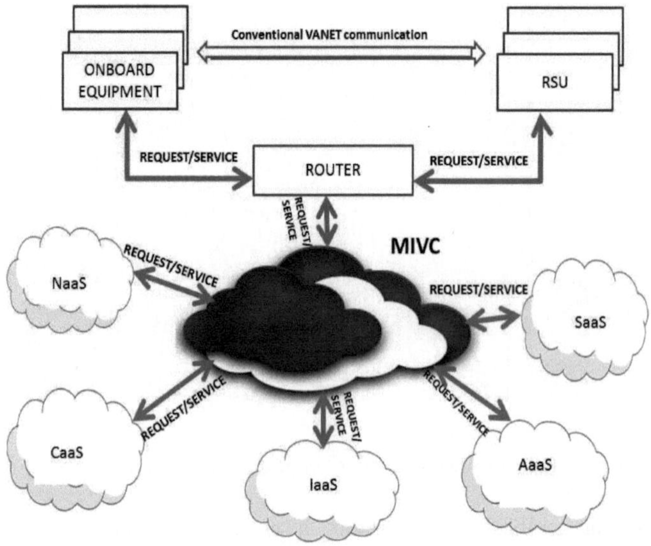

Fig. 2. Data flow within the MIVC

In the Figure 2, communication among vehicular nodes, RSUs and the centralized MIVC are shown. Request for a particular cloud service are sent through a router to the MIVC which would determine which service provide should be responsible for providing the requested service. After the request is processed the results are sent back to the MIVC to be routed to the vehicular node that made the service request.

The interconnection of clouds that is the ability of cloud service providers to exchange data and information among themselves to build a centralized cloud network provides an added advantage to VANET communication. Routing of emergency information can be disseminated very fast. Applications that relay on high computation such as weather prediction and relative speed of nodes can be computed within the central cloud and results sent to individual VANET nodes. This will take the high computation task from the individual vehicular nodes which may not have the needed processing and storage capability.

With one of the main idea behind cloud computing being the full utilization of computer resources, the interconnection of cloud from different service providers would ensure that cloud service providers that have fewer clients but have a relatively large resource base, can rent their resources to other cloud providers at a fee.

Figure 3 shows an image of our proposed MIVC. Here cloud services are easily available on the road unlike in Figure1 where only a section of vehicles may benefit from vehicular cloud formed by nodes in just a part of the road The various clouds are also interconnected to share information about blacklisted nodes. This would ensure that drivers who break traffic regulations at one section of the road or city can easily be arrested anywhere within a shorter time.

Fig. 3. Image showing our proposed MICV

The MIVC would be able to track movements of vehicles within the municipality; this would present an opportunity for the development of new applications such as accurate commercial vehicle schedule and location specific advertisement opportunities.

6.1 Operation of the Proposed MIVC

As nodes enter the metropolitan area termed the MIVC zone, the details of the vehicular nodes are captured with the help of road side units which are connected to the MIVC and equipped sensors. Details such as the vehicle registration number which would serve as the main form of public identification (key) are then recorded in the central database.

With the captured public identity the vehicle is then authenticated as a genuinely registered vehicle and has not been black listed. If authentication fails the law enforcement agencies are quickly informed by sending information from the central cloud system for the vehicle to be subjected to further checks.

If authorization is passed, however, the vehicle automatically gets access to the MIVC and is then assigned a valid certificate (VC) and immediately provided with the services and billing as well as service level agreement (SLA) of different cloud service providers that are within transmission range of the node.

A node that accepts some level of service enters the full MIVC zone whiles nodes that do not accept any of the offers provided by the various cloud service providers in its transmission range joins the partial or limited MIVC zone. In such instance the node may enjoy limited service which could be just the ability receive and retransmit communication data within the VANET environment using the MIVC infrastructure. A flow chart of the architecture of the MIVC is presented below in Figure 4.

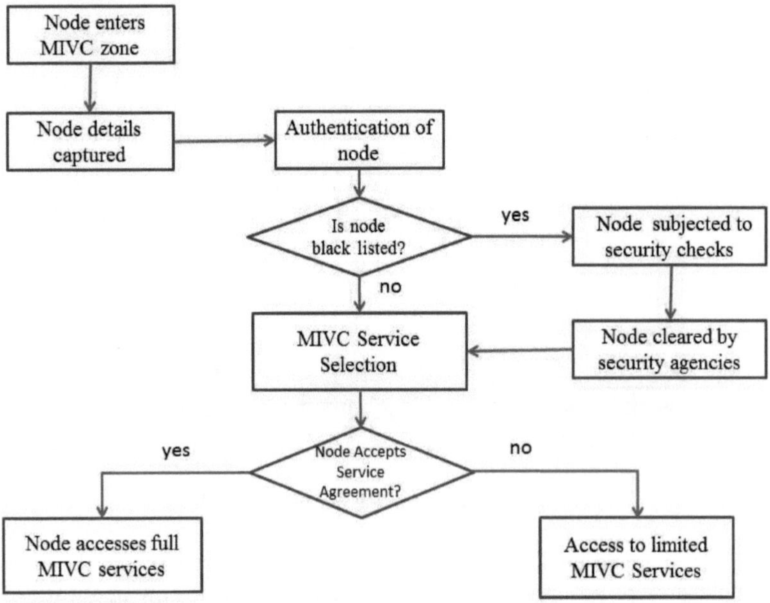

Fig. 4. Architecture of the MIVC

6.2 Authentication as a Service (AaaS) to Enhance Secured Communication in VANET

A novel cloud service, Authentication as a service (AaaS) that is enabled by the MIVC system is proposed. The IEEE 1609.2 standards [6], proposes a Certificate Authority (CA) which is used to authenticate vehicular nodes and issue Validity Certificates. One of the main security challenges in VANET is the distribution of Certificate Renovation List (CRL) explained earlier. To overcome this, the MIVC would provide AaaS to authenticate nodes in a given geographical area. All nodes that enter the MIVC zone are authenticated and given a special ID. Messages that are sent by these nodes are again verified by the MIVC to prevent the verified nodes from sending malicious messages. Nodes that do not have such IDs cannot transmit data to other vehicles. Malicious nodes get their IDs revoked and are prevented from transmitting any information within the network. Here there is no need to send CRL or LRL messages to vehicular nodes or RSUs in the network.

The Authentication as a Service (AaaS) would also be beneficial to traffic law enforcement agencies in that it would ensure that every node on the road network has the needed authorization. AaaS is very significant as it would ensure the possible elimination of malicious nodes in vehicular environment. Once nodes are authenticated their activities and type of messages they transmit within the network can be monitored and quickly punished or blacklisted if it is establish that it is involved in any malicious acts. Access control can be efficiently ensured as protocols and policies can be built on the MIVC to assign distinct roles to different class of nodes in the network. For instance security vehicles (nodes) can be assigned the role of sharing information about vehicles that break traffic regulations for further action.

6.3 Economic and Political Issues

One of the challenging issues that need to be addressed in MIVC is the economic and political aspects of the system. The MIVC system would gain universal acceptance if it is economically feasible for both road users and service providers. In our proposed MIVC system, cloud services could be provided solely by using resources that belongs to the service provider or a collaboration of free resources such as processing cycles and storage of vehicles that are on the road or parked at private and commercial car parks which can be pooled and used as part of the MIVC. A hybrid of the two types could also be employed in the proposed system.

The issue that arises here is; are clients who provide resources for cloud service providers in cloud formation to pay for cloud services they enjoy from the MIVC system or not. These and other challenging issues such as pricing, cost to be incurred by vehicular nodes that do not want to access any cloud service within city-wide MIVC needs to be studied for the system to be more feasible economically.

To address issues of privacy of the users of the MVIC, only restricted information would be made available for tracking by the MIVC.

7 Conclusion and Future Works

A novel Metropolitan interconnected Vehicular Cloud (MIVC) system consisting of different cloud providers, administered and coordinated by a single entity usually the municipal or metropolitan assembly is proposed. The MIVC is to provide cloud services in VANET environment as well as overcome some of the security challenges in VANET.

It is worth stating that the proposed idea of the interconnection of cloud services and resources from different cloud providers to the best of our knowledge does not exist in any known literature. This can extended to other traditional cloud systems and other VCloud to ensure efficient and full utilization of cloud resources of different cloud service providers.

Interconnection of cloud services and resources from different cloud providers comes with other issues such as security, privacy and interoperability. In future, these issues would be examined to provide standards and principles to ensure that such challenges are overcome. Simulations of the proposed framework would also be carried out in future works.

References

[1] U.S. Department of Transportation, Intelligent Transportation Systems (ITS) Home, `http://www.its.dot.gov/index.htm` (date retrieved October 12, 2014)

[2] `http://www.asirt.org/KnowBeforeYouGo/RoadSafetyFacts/RoadCrashStatistics/tabid/213/Default.aspx` (date retrieved August 2, 2014)

[3] Dedicated Short range Communications (DSRC) Medium Access Control (MAC) and Physical Layer (PHY) Specifications. US Department of Transportation Standard Specification for Telecommunications and Information Exchange Between Roadside and Vehicle Systems–5GHZ Band. ASTM E2213-03 (August 2003), `http://www.leearmstrong.com/DSRC/DSRCHomeset.htm`

[4] `http://www.etsi.org/technologiesclusters/technologies/intelligent-transport/dsrc` (date retrieved February 2, 2014)

[5] Ezell, S.: Intelligent Transport Systems. The Information Technology and Information Foundation (ITTF) explaining international IT application leadership (January 2010)

[6] Raya, M., Papadimitratos, P., Hubaux, J.-P.: Securing Vehicular Communications. IEEE Wireless Communications Magazine, Special Issue on Inter-Vehicular Communications (October 2006)

[7] Laberteaux, K.P., Haas, J.J., Hu, Y.-C.: Security certificate revocation list distribution for vanet. In: VANET 2008, Proceedings of the Fifth ACM International Workshop on VehiculAr Inter-NETworking, New York, USA, pp. 88–89 (2008)

[8] Yan, G., Wen, D., Olariu, S., Weigle, M.C.: Security challenges in Vehicular cloud Computing. IEEE Transactions on Intelligent Transportation Systems 14(1) (March 2013)

[9] Mathew, M.E., Arun Raj Kumar, P.: Threat Analysis and Defence Mechanisms in VANET. International Journal of Advanced Research in Computer Science and Software Engineering 3(1) (January 2013)

[10] NIST Special Publication 800-146, Cloud Computing Synopsis and..Recommendations, `http://csrc.nist.gov/publications/nistpubs/800-146/sp800-146.pdf`

[11] Papadimitratos, P., Buttyan, L., Holczer, T., Schoch, E., Freudiger, J., Raya, M., Ma, Z., Kargl, F., Kung, A., Hubaux, J.-P.: Secure vehicular communications: design and architecture. IEEE Communications Magazine 46(11), 100–109 (2008)

[12] Leinmuller, T., Schoch, E., Kargl, F.: Position Verification Approaches for Vehicular Ad-Hoc Networks. IEEE Wireless Communications (October 2006)

[13] Rawat, A., Sharma, S., Sushil, R.: VANET: Security Attacks and Its Possible Solutions. Journal of Information and Operations Management 3(1), 301–304 (2012) ISSN: 0976–7754 & E-ISSN: 0976–7762

[14] IEEE Trial-Use Standard for Wireless Access in Vehicular Environments - Security Services for Applications and Management Messages, IEEE Std 1609.2-2006, 0_1,105 (2006)

[15] Samara, G.: Certificate Revocation Management in VANET. International Journal of Cyber-Security and Digital Forensics (IJCSDF) 1(2), 115–121 (2012)

[16] IDC, Worldwide and Regional Public IT Cloud Services 2010-2014 Forecast, Doc # 223549 (June 2010)
[17] http://www.att.com/gen/pressroom?pid=19057&cdvn=news&newsarticleid=31603 (date retrieved December 2, 2013)
[18] http://www.prnewswire.com/news-releases/turkcell-superonline-provides-turkeys-first-personal-cloud-service-superdepo-at-anupload-speed-of-100-mbps-183467071.html (date retrieved July 12, 2013)
[19] Brandao, A.S., Sasaki, A.S., Castelano, C.R., Cruz, R.R., Carelli, R.: Autonomous Navigation with Obstacle Avoidance for a Car-Like Robot. In: Robotics Symposium and Latin American Robotics Symposium (SBR-LARS), Brazilian, October 16-19, pp. 156–161 (2012)
[20] Hussain, R., Son, J., Eun, H., Kim, S., Oh, H.: Rethinking Vehicular Communications: Merging VANET with cloud computing. In: 2012 IEEE 4th International Conference on Cloud Computing Technology and Science (CloudCom), December 3-6, pp. 606–609 (2012)
[21] Al Mamun, M.A., Anam, K., Onik, M.F.A., Esfar-E-Alam, A.M.: Deployment of Cloud Computing into VANET to create Ad-Hoc Cloud Network Architecture. In: Proceedings of the World Congress on Engineering and Computer Science, WCECS 2012, vol. 1, San Francisco, USA (October 2012)
[22] Anda, J., LeBrun, J., Ghosal, D., Chuah, C.-N., Zhang, M.: VGrid: vehicular adhoc networking and computing grid for intelligent traffic control. In: 2005 IEEE 61st Vehicular Technology Conference, VTC 2005-Spring, May 30-June 1, vol. 5, pp. 2905–2909 (2005)
[23] Abuelela, M., Olariu, S.: Taking VANET to the clouds. In: Proceedings of the 8th International Conference on Advances in Mobile Computing and Multimedia (MoMM 2010), pp. 6–13. ACM, New York (2010)
[24] Whaiduzzaman, M., Sookhak, M., Gani, A.: A survey of vehicular cloud computing. Journal of Network and Computer Applications (2013)

Design of Dual Band H-Shaped Antenna for DCS and WLAN Applications

K. ThanaPakkiam[1,*], K. Baskaran[1], and J.S. Mandeep[2]

[1] Department of Electrical Electronics Engineering,
Faculty of Engineering Technology Infrastructure,
Infrastructure University Kuala Lumpur, Selangor, Malaysia
[2] Department of Electrical, Electronics & System,
Faculty of Engineering and Built Environment, National University of Malaysia,
Bangi, Selangor, Malaysia
{thanapakkiam,baskaran}@iukl.edu.my,
mandeep@eng.ukm.my

Abstract. This paper presents the performance of a dual-band H-shaped patch antenna for DCS and WLAN applications. The proposed antenna employs a microstrip feed line and a FR4 substrate on which an H-shaped slot is cut using the etching technique. This structure radiates for Digital Communication System DCS (1.710-1.880) GHz and for Local Area Network WLAN (5.15 -5.35) GHz. A commercially available software CST Microwave Studio simulation showed that the proposed design exhibits a return loss of 15.36 dB and 23.493 dB at 1.81 GHz and 5.295 GHz respectively. Further, for each there was a power gain of 2.45 dBi and 2.1 dBi, and the VSWR was less than 2 for both bands while efficiency is between 80 to 90%. The proposed antenna could possibly be employed in a modern communication system that has constraints in size and weight.

Keywords: H-shaped antenna, dual band antenna, DCS and WLAN antenna, H-shaped patch antenna.

1 Introduction

Tremendous growth in wireless markets demand product that are capable of providing multiple radio services within a single device. The desired antennas will be a low profile and smaller size with multiband characteristics. Besides, simpler fabrication methodology is also preferable. Microstrip patch configuration is a potential candidate that can be considered for the current radio designs due to low fabrication cost, low profile and ability to integrate with other electronic devices. Various small printed monopole antennas are published in the literature for wireless applications. Further several designs of microstrip patch antennas are reported. A tri-band H- shaped microstrip patch antenna for DCS and WLAN applications [1]. In [2], [3], [4] multiband antennas by etching two slots with different lengths on a wideband monopole structure. In [5], trapezoidal ground is used to achieve multiband resonance for WLAN and

* Corresponding author.

K.J. Kim and N. Wattanapongsakorn (eds.), *Mobile and Wireless Technology 2015*,
Lecture Notes in Electrical Engineering 310, DOI: 10.1007/978-3-662-47669-7_2

WIMAX applications. There are numerous and well known method is used to increase the bandwidth of antenna [6]. Many researcher have proposed diverse shapes of microstrip patch antennas for distinct applications with different feeding techniques [7], [8]. A number of slot shapes have been designed and proposed e.g. E-shaped [9], H-shaped [8], C- shaped [10] and U-shaped [11].

In this article, a simple H-shaped monopole patch with defected ground structure is proposed to achieve the dual-band operation performance. This structure is smaller in size and simpler to fabricate. By using the two different resonant frequencies, the proposed antenna can generate two resonant modes to cover two desired bands for DCS and WLAN applications. This article is organized as follow. Section 2 presents the geometry and the design methodology of the proposed antenna in detail. Simulation and experimental results are presented in Section 3. In Section 4 a brief conclusion is outlined.

2 Antenna Structure

Figure 1 shows the structure and dimensions of the proposed antenna of 35 x 50 mm^2 with conductor fabricated on an inexpensive FR4 substrate having the dielectric constant of 4.3 and the substrate thickness of 1.6 mm. The antenna's shape and its dimensions were first searched by using the commercially available software CST Microwave Studio [12] and then the optimal dimensions were determined from experimental adjustments.

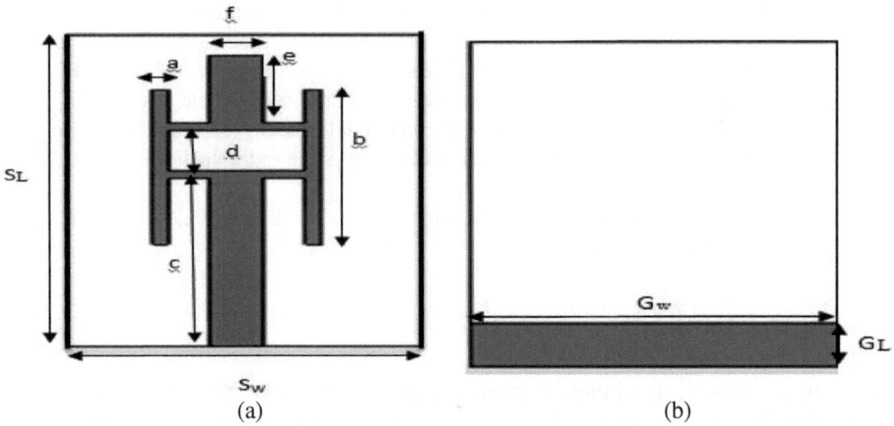

Fig. 1. Geometry and dimensions of proposed antenna (a) front view and (b) back view

The basic antenna structure consists of a rectangular patch, a feed – line, and a ground plane. The radiating patch is connected to a feed line. The proposed antenna is connected to a 50 Ω SMA connector for signal transmission. The top view of proposed antenna is shown in Figure 1(a) and the back view is shown in Figure 1 (b). Feed line is fixed as 4.5 x 20 mm^2 to meet the resonant mode with a ground plane dimension of 35 x 6 mm^2. The radiating elements of H-shaped and feed line is printed on top of the substrate, while the partial ground is printed on the other side of the substrate. The optimized value of the parameters of the proposed antenna are listed in Table 1.

Table 1. Optimized dimensions of the geometric parameters

Parameters		Dimensions in mm
Substrate	- S_W	35
	- S_L	50
Ground	- G_W	35
	- G_L	6
Feed line	- c, e, f	20, 10, 3.4
patch	- a, b	1 , 23
Patch	- d	6
Copper thickness		0.035

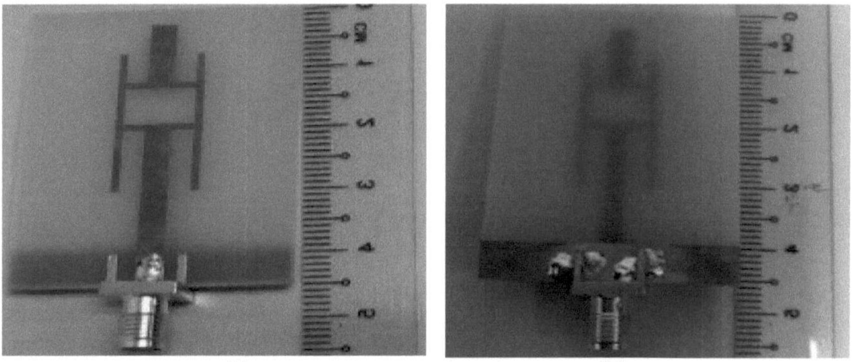

(a) (b)

Fig. 2. Fabricated antenna (a) Front view and (b) back view of the proposed antenna

3 Simulation and Measurement Results

To verify the proposed antenna, an experimental prototype as shown in Figure 2 (a) front view and (b) back view, is fabricated and measured. All s-parameter measurements of the antennas fabricated and presented in this project were done on a Rhode & Schwarz Vector Network Analyzer (Model ZVL 30) that was available at IUKL University's laboratory. Radiation pattern, gain and efficiency measurements were carried out at Atenlab's (Taiwan) antenna measurement facility. The equipment used for the measurement at Atenlab's is shown in Figure 3.

The measured and simulated S11 against frequency for the presented antenna are plotted in Figure 4. From the results, both resonant modes achieved a -10 dB for simulations at 1.826 GHz and 5.315 GHz are successfully excited. Obviously, the antenna can operate for the first resonant mode at about 1.81 GHz which can be used for DCS and the second resonant mode at about 5.315 GHz applied for WLAN operations. We measured the data with the simulated results obtained from the electromagnetic solver. Good agreement is seen over the lower operating band, while a reasonable agreement with frequency shift for the upper band is also seen. The discrepancies between the measured and simulated return loss may be due to the frequency response of the substrate permittivity and the exactly calculated feeding

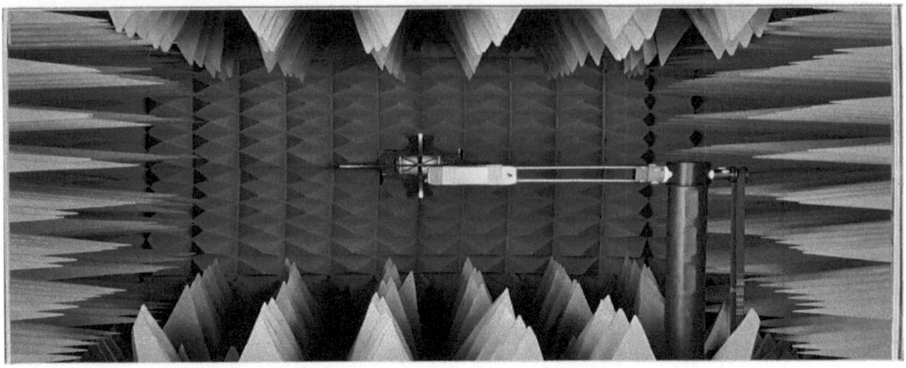

Fig. 3. Proposed antenna under test for measurement

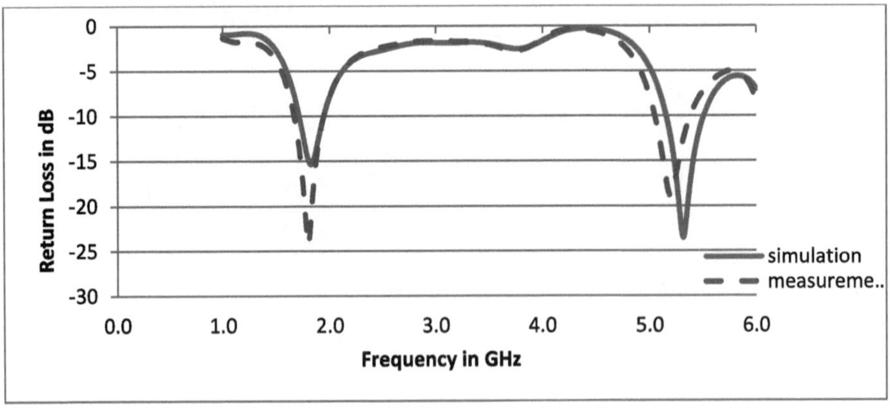

Fig. 4. Simulated and measured return loss of the proposed antenna

structure. An important feature of the proposed antenna is the capability of impedance matching at both operating frequencies using single feed line and partial ground. A parametric study has been performed to enhanced the performances by evaluating the effects of the various patches and partial ground on the reflection coefficients.

3.1 Effect of Partial Ground Height 'G_L' and Width 'G_w'

Figure 5 presents the simulated results of the return loss against frequency with different height of partial ground 'Gl' for the proposed antenna. Referring to Figure 5, it is observed that the partial ground height has a large impact on the impedance band width for both upper and lower bands with the partial ground height varying from 4 mm to 7 mm. However for 6 mm height, it shows good performances. The ground height 'Gl' is the major contributor for the lower frequency at 1.81 GHz. The effect of partial ground width is also investigated. The simulated return losses of different

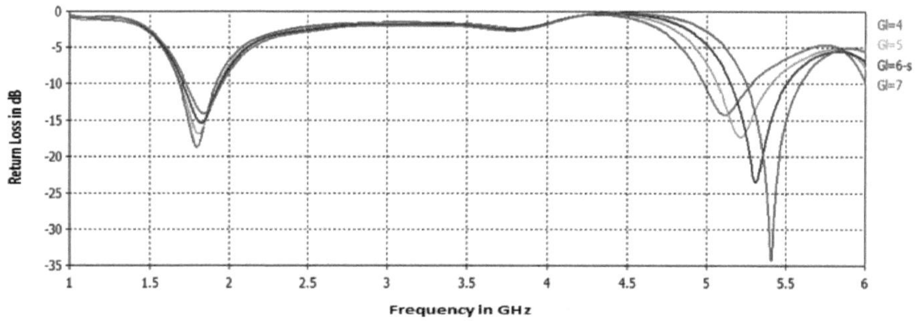

Fig. 5. Reflection coefficients for various partial ground height 'G_L'

ground width of 'Gw' are shown in Figure 6. From the results, the low frequency band located at 1.826 GHz was not affected by these changes. However, the upper frequency moved from 5.314 GHz to 5.45 GHz and this movement showed thats the presence of selected width has generated higher resonant frequency.

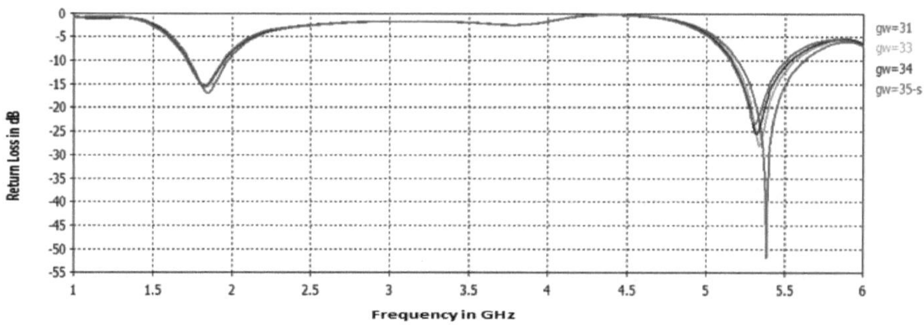

Fig. 6. Reflection coefficients for various partial ground width 'G_w'

3.2 Effect of Different Patch Elements 'e' and 'b'

Another parameter investigated is the effect of patch 'e' as shown in Figure 7. It is noticed that the height of patch element 'e' has a large impact on the impedance bandwidth of the upper band with the height varying from 8 mm, while the lower band almost keeps unchanged. For 10 mm height, the low frequency band located at 1.826 GHz was not affected by these changes. However, the high frequency band moved from 5.41 GHz to 5.315 GHz and also showed improvement on the return loss. The effect of height 'b' of patch element is studied on the impedance matching for the proposed antenna as shown in Figure 8. From the results, it is observed that the height of patch element 'b' has a large impact on the impedance bandwidth of the upper band with height varying form 21 mm to 27 mm. The lower band has minimum effect on return loss, with an increase of height from 21 mm to 25 mm, the resonant mode decreased but the return loss increased. Therefore the suitable height is fixed to 16 mm, for WLAN applications.

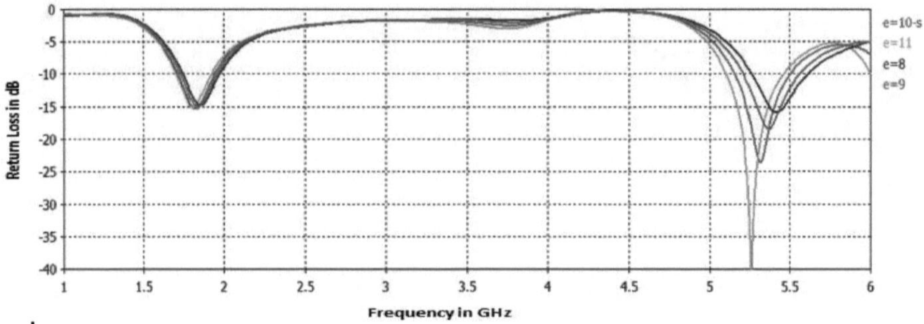

Fig. 7. Reflection coefficients for various patch elements height 'e'

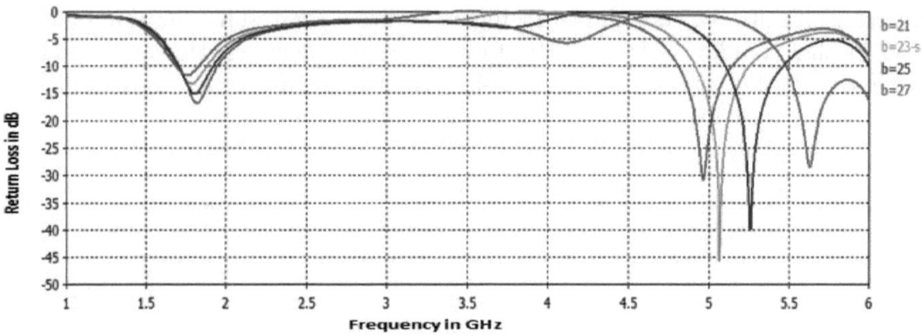

Fig. 8. Reflection coefficients for various patch elements height 'b'

3.3 Simulated Surface Current Distribution and Density

The current density at 1.81 GHz and 5.314 GHz is illustrated in Figure 9 (a) and (b) respectively. Similarly, Figure 10 (a) and (b), depict the excited surface current distributions of the proposed antenna, including the current flow on H-shaped area of the patch. It can be seen that the surface current distribution of 1.81 GHz is concentrated at feed line while that of 5.314 GHz is concentrated at around the H-shaped element. Arrows shows the direction of the current distribution. It can be observed for the Figures that the current intensely flows at the edge of the slots especially near the feeding probe of the patch. However, the current is uniformly distributed elsewhere. Figure 11 shows the simulated and measured radiation patterns E-plane and H-plane for its resonant frequencies of 1.81 GHz and 5.314 GHz. Both resonant frequencies have figure-eight bidirectional patterns in E-plane and omnidirectional patterns in H- plane.

Figure 12, shows the simulated and measured power gain at 1.81 GHz and (b) 5.314 GHz are 2.45 dBi and 2.41 dBi respectively. Also it is observed from the VSWR plot in Figure 13 that there was satisfying good agreement for the resonant frequencies which is less than 2 throughout. The antenna has a maximum of 89.8% radiation efficiency with an average of 72.3%. Due to the fabricated dielectrics substrate with modified partial ground plane, the dielectric loss was high, which affected the efficiency as shown in Figure 14.

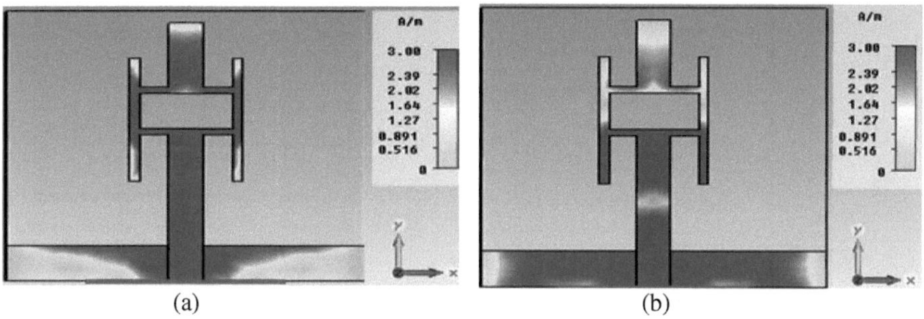

Fig. 9. Current density of the proposed antenna at (a) 1.81 GHz and (b) 5.314 GHz

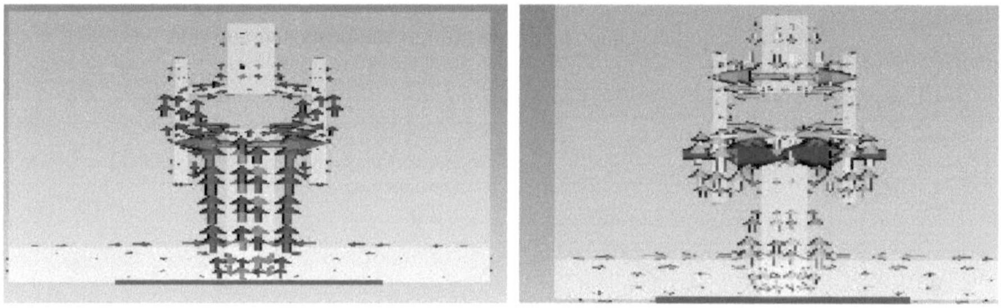

Fig. 10. Current distribution of the proposed antenna at (a) 1.81 GHz and (b) 5.314 GHz

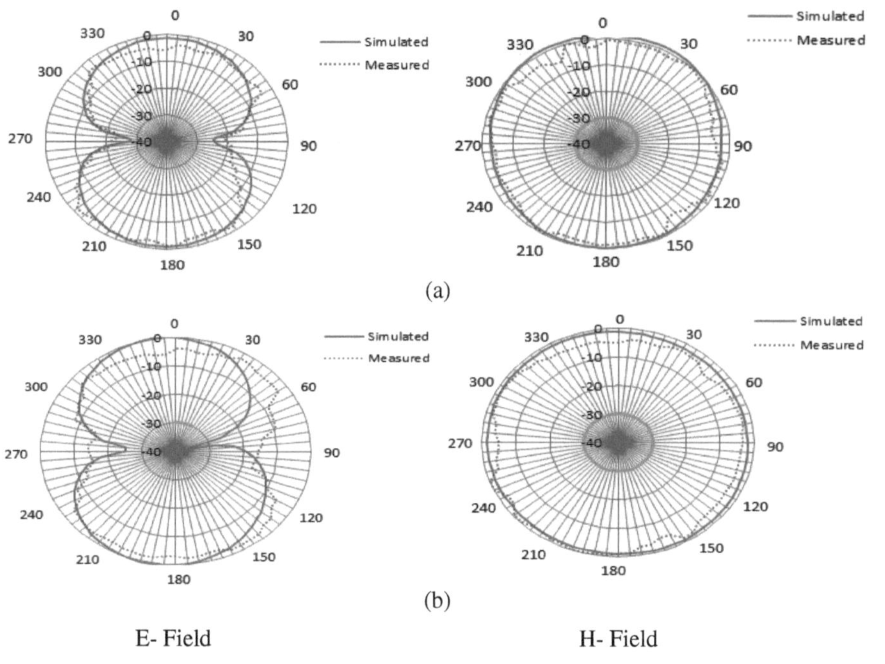

Fig. 11. Simulated and measured radiation pattern of the proposed antenna at (a) 1.81 GHz and (b) 5.314 GHz

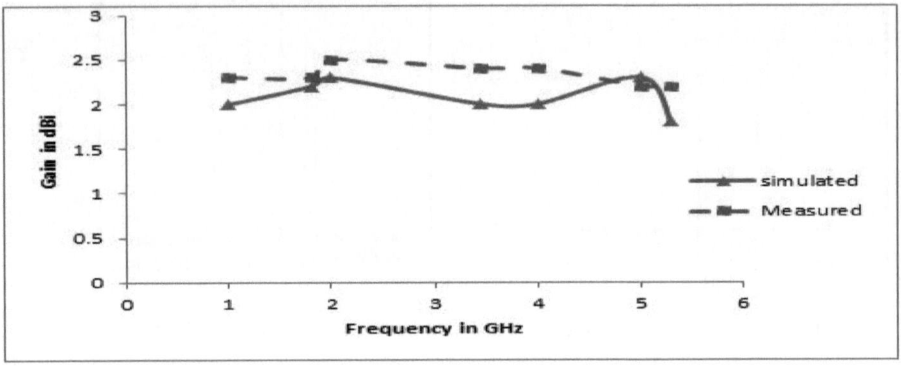

Fig. 12. Simulated and measured gain of the proposed antenna

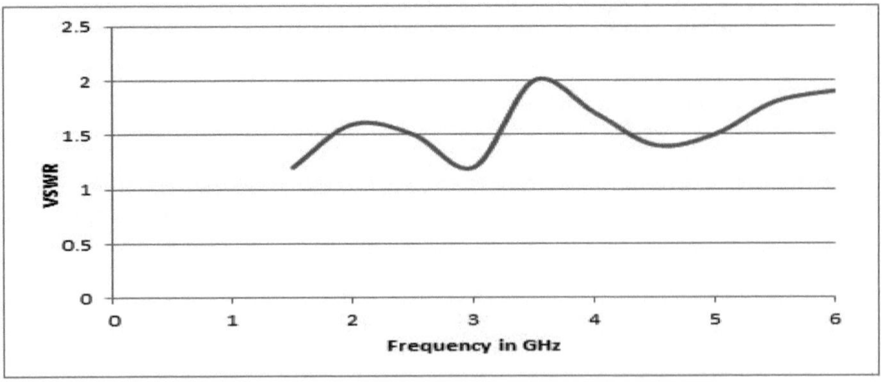

Fig. 13. Simulated Voltage Standing Wave Ratio (VSWR) plot

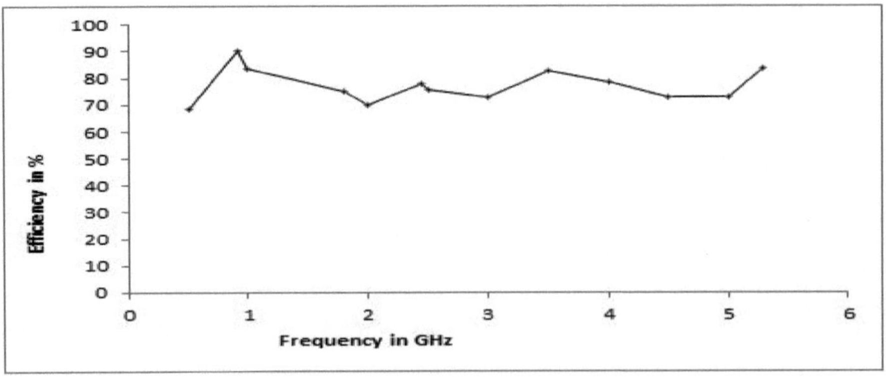

Fig. 14. Calculated radiation efficiency

4 Conclusion

The design of dual band H-shaped antenna for DCS and WLAN applications have been developed and implemented successfully. The antenna resonates at 1.81 GHz (DCS) and 5 .314 GHz (WLAN) with good return loss and VSWR of less than 2 which can be suitable for cellular communications. The maximum achievable gain of the antenna is 2.45 dBi. Additionally, the radiation patterns of the proposed antenna are bidirectional in E-plane and omnidirectional in H-plane at 1.81 GHz and 5 .314 GHz. The proposed antenna has a very simple structure, which makes the design simpler and enables easy fabrication.

References

1. Afzal, W., Rafique, U., Ahmad, M.M., Khan, M.A., Mughal, F.A.: A tri- band H-shaped microstrip patch antenna for DCS and WLAN application. In: 19th Int. Conf. Microwave Radar and Wireless Communications (MIKON), vol. 1, pp. 256–258 (2012)
2. Cai, L.Y., Zeng, G., Yang, H.C.: Compact triple band antenna for Bluetooth/WiMax/WLAN applications. In: Proceeding of International Symposium on Signals, Systems and Electronics, Nanjing, China, September 17-20, pp. 1–4 (2010)
3. Krishna, D.D., Gopalakrishna, M., Aanandan, C.K.: A CPW-fed triple band monopole antenna for WiMAX and WLAN applications. In: Proceeding of the 38th Eur. Microwave Conf., Amsterdam, Netherland, pp. 897–900 (October 2008)
4. Liu, H.W., Ku, C.H., Yang, C.F.: Novel CPW-fed planar monopole antenna for WiMax /WLAN applications. IEEE Antennas Wireless Propag. Lett. 9, 240–243 (2010)
5. Thomas, K.G., Sreenivasan, M.: Compact triple band antenna for WLAN/WiMAX applications. Electronics Lett. 45(16), 811–813 (2009)
6. Chang, E., Long, S., Richards, W.F.: An Experimental investigation of electrically thick rectangular microstrip. IEEE Transactions on Antenna and Propagation 34(6), 767–772 (1986)
7. Siew-Bee, Y., Ning, C.Z.: Microstrip Patch Antenna with Enhanced Gain by Partial Substrate Removal. IEEE Transactions on Antenna and Propagation 58(9), 2811–2816 (2010)
8. Lu, W.K., Hao-Chun, T., Wern, C.T.: Broadband dual–polarized aperture-coupled patch antennas with modified H-shaped coupling slots. IEEE Transactions on Antenna and Propagation 50(2), 191 (2002)
9. Izadi Hoseini, O., Mehrparvar, M.: A Compact microstrip slot antenna with novel E-shaped coupling aperture. In: 5th International Symposium on Telecommunications (IST), pp. 110–114 (2010)
10. Padhi, S.K., Karmaker Sr., N.C., Law, C.L., Adiya Sr., S.: A dual polarized aperture circular patch antenna using a C-shaped coupling slot. IEEE Transactions on Antenna and Propagation 51(12), 3295–3298 (2003)
11. Guo, Y.X., Luk, K.M., Lee, K.F., Chow, Y.L.: Double U-slot rectangular patch antenna. Electronics Lett. 34(19), 1805–1806 (1998)
12. CST Microwave Studio Electromagnetic Field Simulation Software, Computer Simulation Technology, Darmstadt, Germany

Secrecy Rate of the Amplify-and-Forward Relay Wire-Tap Channel in Rayleigh Fading

Cuong Dang[1], Leonardo J. Rodríguez[2], Duc-Anh Le[1], Forrest Sheng Bao[1], and Nghi H. Tran[1]

[1] Department of Electrical & Computer Engineering, University of Akron, Akron, OH USA
[2] Department of Electrical & Computer Engineering, McGill University, Montréal, QC Canada

Abstract. Benefiting from information-theoretic studies in cooperative relaying communications, relaying strategies have recently received considerable attention in the context of physical-layer security over wireless networks. In this paper, we investigate the secrecy rate of a relay wire-tap Rayleigh fading channel in which a source wishes to communicate securely to a destination in the presence of an eavesdropper and under the aid of an amplify-and-forward (AF) relay. Due to the presence of fading, it is a very challenging task to obtain the secrecy rate in closed-form. To overcome this issue, we first propose a novel method to calculate the expectation of an exponentially distributed random variable using the exponential integral function. By exploiting this calculation, we then establish the average secrecy rate of the considered AF relay channel in closed-form. Numerical results show that the proposed formula provides an accurate and effective way to obtain the secrecy rate without the need of time-consuming Monte Carlo simulations.

Keywords: PHY security, relay channel, secrecy rate, wire-tap channel.

1 Introduction

Relaying techniques, in which a relay node assists a source node to communicate to a destination node, have received considerable attention from both industry and academia. The deployment of relays has been shown to increase the range and reliability of wireless networks in a cost-effective manner (see for example [7-9,16] and reference within). In general, relaying strategies can be categorized as decode-and-forward (DF), compress-and-forward (CF) and amplify-and-forward (AF) of which the AF scheme is the simplest to implement since the relay only needs to scale and retransmit the received signal. Among different AF schemes, the dual-hop AF system without direct link from the source to destination has attracted considerable efforts in the literature [4,10,12].

Due to the broadcast nature of the wireless channel, it is also important to transmit information in a confidential manner in the presence of eavesdroppers. In this context, physical layer (PHY) security metrics based on information theory such as secrecy

K.J. Kim and N. Wattanapongsakorn (eds.), *Mobile and Wireless Technology 2015*,
Lecture Notes in Electrical Engineering 310, DOI: 10.1007/978-3-662-47669-7_3

rate and secrecy capacity of which the theoretical foundation were laid by Wyner in [18] have been gaining increasing research attention.

Recently, relaying has also been shown to be a very attractive solution to enhance the secrecy rate of wireless networks under the context of PHY security [1–3,5,6,11,15,17,19]. For instance, as in cooperative relay communications, relay nodes can be deployed to retransmit an amplified version of the signal received from the source with a suitable power amplification coefficient in AF mode [11]. The trusted relay can also transmit a weighted version of the decoded signal in DF mode, or forward a compressed copy of the received signal in CF mode. These trusted relays can also be used to steer the information vector away from the eavesdropper and in the direction of the intended destination through distributed beamforming [2,3,6,17]. Alternatively, relays can also be used as jammers to degrade the signal-to-noise ratio (SNR) at the eavesdropper without degrading the SNR at the destination [1,3,5,15,19].

The benefits offered by cooperative relaying to enhance the security at the physical layer are significant. However, dealing with secrecy rate and secrecy capacity to understand the fundamentals and insights of PHY security in a wireless relay network still poses many challenges. In particular, the presence of fading and various relaying strategies makes it difficult to calculate the secrecy rate in an effective manner. To our knowledge, such calculation usually relies on lengthy Monte Carlo simulations.

Motivated by the above observations, in this paper, we study the secrecy rate of a relay wire-tap Rayleigh fading channel in which a source wishes to communicate securely to a destination in the presence of an eavesdropper and under the aid of an amplify-and-forward (AF) relay. To overcome the complexity from the stochastic nature of fading, we first propose a novel method to calculate the expectation of an exponentially distributed random variable using the exponential integral function. Using this result, the average secrecy rate of the considered AF relay channel is then established in closed-form. Numerical results in various network configurations demonstrate that the proposed method provides an accurate and effective way to calculate the secrecy rate without the need of time-consuming Monte Carlo simulations.

2 System Model

As shown in Fig. 1, the considered half-duplex (HD) AF relay system consists of four single-antenna nodes: a source node S, a relay node R, a destination node D, and an eavesdropper node E. The transmission is carried out in frames composed of two consecutive unit-time phases. In the first phase of a given frame i, S transmits the signal x_i to the relay R. The signal received at R then can be written as $r_i = \sqrt{P_s} h_{sr}^{(i)} x_i + n_{r,i}$, where P_s is the power transmitted by S in the first phase with $E[|x_i|^2] = 1$; $h_{sr}^{(i)}$ is the S-R complex channel gain at frame i; and $n_{r,i} \sim \mathcal{CN}(0, N_r)$ is the zero-mean circularly Gaussian noise at R.

In the second phase of frame i, R amplifies and forwards the symbol received during the first phase to D using an amplification coefficient b. Thus, the signal transmitted by R at the frame i is given by $t_i = br_i$. Then, the signal received at D and overheard at E in the second phase can be written respectively as

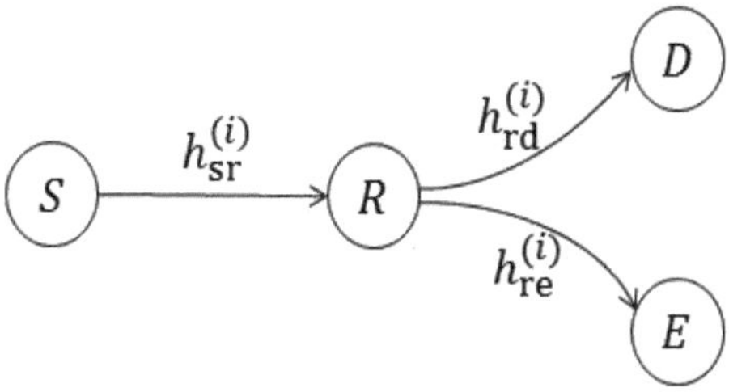

Fig. 1. The HD AF wiretap channel

$$\begin{aligned} y_{d,i} &= \sqrt{P_r} h_{rd}^{(i)} t_i + n_{d,i} = \sqrt{P_s P_r} h_{sr}^{(i)} h_{rd}^{(i)} b x_i + \sqrt{P_r} h_{rd}^{(i)} b n_{r,i} + n_{d,i}, \\ y_{e,i} &= \sqrt{P_r} h_{re}^{(i)} t_i + n_{e,i} = \sqrt{P_s P_r} h_{sr}^{(i)} h_{re}^{(i)} b x_i + \sqrt{P_r} h_{re}^{(i)} b n_{r,i} + n_{e,i}, \end{aligned} \tag{1}$$

where P_r is the power transmitted by R in the second phase; $h_{rd}^{(i)}$ and $h_{re}^{(i)}$ are the R-D and R-E complex channel gains at frame i; $n_{d,i} \sim \mathcal{CN}(0, N_d)$ and $n_{e,i} \sim \mathcal{CN}(0, N_e)$ are the noise samples at D and E respectively. To keep the power constraint of P_r at R, we assumed that the relay has instantaneous knowledge of the S-R link and applies a variable gain amplification coefficient as $b = \sqrt{1/\left(|h_{sr}^{(i)}|^2 P_s + N_0\right)}$. D and E are also assumed to have instantaneous knowledge of all channel gains $h_i = [h_{sr}^{(i)}, h_{rd}^{(i)}, h_{re}^{(i)}]$, whereas the source S only has statistical knowledge. Furthermore, the channel gains are independent zero-mean complex circular Gaussian with arbitrary variances as $h_{sr}^{(i)} \sim \mathcal{CN}(0, \phi_{sr})$, $h_{rd}^{(i)} \sim \mathcal{CN}(0, \phi_{rd})$, $h_{re}^{(i)} \sim \mathcal{CN}(0, \phi_{re})$. The noise variances at all nodes are set to the same value, i.e., $N_r = N_e = N_d = N_0$. Assuming Gaussian codebooks at S, the achievable rate at the destination D and the eavesdropper E conditioned on a realization of h_i can now be written (in b/s/Hz) from (1) and b as

$$\begin{aligned} I(x_i, y_{d,i} | h_i) &= \frac{1}{2} \log\left(1 + \frac{\gamma_{sr}\gamma_{rd}}{\gamma_{sr} + \gamma_{rd} + 1}\right), \\ I(x_i, y_{e,i} | h_i) &= \frac{1}{2} \log\left(1 + \frac{\gamma_{sr}\gamma_{re}}{\gamma_{sr} + \gamma_{re} + 1}\right), \end{aligned} \tag{2}$$

where $\gamma_{sr} = |h_{sr}^{(i)}|^2 P_s / N_0$, $\gamma_{rd} = |h_{rd}^{(i)}|^2 P_r / N_0$, $\gamma_{re} = |h_{re}^{(i)}|^2 P_r / N_0$ and $\log(\cdot)$ is the base-2 logarithm. By taking the average over fading realizations, the average achievable secrecy rate can then be expressed as:

$$R_s = E_h[I(x_i, y_{d,i} | h_i) - I(x_i, y_{e,i} | h_i)]^+, \tag{3}$$

where $[x]^+ = \max\{0, x\}$ and E_h denotes the expectation over all the realizations of fading gains of *S-R*, *R-D* and *R-E* channels. The objective of this paper is to find a closed-form formula for the above secrecy rate.

3 Closed-Form Secrecy Rate

As can be seen from (3), it is not an easy task to obtain R_s in closed-form. Due to the presence of fading, the secrecy rate involves a triple-integral and numerically calculating it with high accuracy is very cumbersome. As an alternative, in the following, we demonstrate that a closed-form expression of the secrecy rate can be established. To this end, we first express R_s in (3) as $R_s = E_h[1/2\times\log(g(h_i))]^+$, where $g(h_i)$ is given by

$$g(h_i) = \left(1 + \frac{\gamma_{sr}\gamma_{rd}}{\gamma_{sr} + \gamma_{rd} + 1}\right) \Bigg/ \left(1 + \frac{\gamma_{sr}\gamma_{re}}{\gamma_{sr} + \gamma_{re} + 1}\right) = \frac{(\gamma_{rd}+1)(\gamma_{sr}+\gamma_{re}+1)}{(\gamma_{re}+1)(\gamma_{sr}+\gamma_{rd}+1)}. \quad (4)$$

Therefore, from (3), the secrecy rate can be factored as

$$\begin{aligned} R_s = 1/2\times\big\{ & E_{h_{rd}}[\log(1+\gamma_{rd})] + E_{h_{sr},h_{re}}\left[\log(1+\gamma_{sr}+\gamma_{re})\right] \\ & - E_{h_{re}}\left[\log(1+\gamma_{re})\right] - E_{h_{sr},h_{rd}}[\log(1+\gamma_{sr}+\gamma_{rd})]\big\}. \end{aligned} \quad (5)$$

To further examine the expectations in (5), we first have the following lemma with regards to the exponential integral.

Lemma 1. *Let ω_1 and ω_2 be independent exponentially distributed random variables with means ϕ_1 and ϕ_2, respectively. Define $\mathcal{J}(x) = \exp(x)E_1(x)$ (6), where $E_1(.)$ is the exponential integral*

$$E_1(x) = \int_x^\infty \frac{e^{-u}}{u} du = -\left(\gamma + \ln(x) + \sum_{n=1}^{\infty} \frac{(-1)^n x^n}{n!n}\right),$$

and γ is the Euler constant. Then,

$$E_{\omega_1}[\ln(1+\omega_1)] = \mathcal{J}(1/\phi_1), \quad (7a)$$

$$E_{\omega_1,\omega_2}[\ln(1+\omega_1+\omega_2)] = \begin{cases} 1 + \left(1 - \dfrac{1}{\phi_1}\right)\mathcal{J}(\dfrac{1}{\phi_1}), \phi_1 = \phi_2 \\ \dfrac{\phi_1\mathcal{J}(1/\phi_1) - \phi_2\mathcal{J}(1/\phi_2)}{\phi_1 - \phi_2}, \phi_1 \neq \phi_2. \end{cases} \quad (7b)$$

Proof. (7a) can be written as the rate (in nats/s/Hz) of a single-input-single-output system with instantaneous SNR ω_1, which is given by [14, eq. (15.26)]. Moreover, (7b) can be written as the rate of a two-branch maximum-ratio combiner with SNRs ω_1 and ω_2. This rate is given by [14, eq. (15.33)] when the average SNRs are equal ($\phi_1 = \phi_2$) and by [13] for unequal average SNRs ($\phi_1 \neq \phi_2$).

Now let $\overline{\gamma_{sr}} = E\{\gamma_{sr}\}=P_s\phi_{sr} / N_0$, $\overline{\gamma_{rd}} = E\{\gamma_{rd}\}=P_r\phi_{rd} / N_0$, and $\overline{\gamma_{re}} = E\{\gamma_{re}\}=P_r\phi_{re} / N_0$. From 7(a) and 7(b) in Lemma 1, the expectation in (5) can be obtained in closed-form as follows:

$$R_s = \frac{1}{2}\times\begin{cases} \frac{1}{\ln 2}\left[\frac{1}{\overline{\gamma_{rd}}}J\left(\frac{1}{\overline{\gamma_{rd}}}\right)-\frac{1}{\overline{\gamma_{re}}}J\left(\frac{1}{\overline{\gamma_{re}}}\right)\right], \text{if } \begin{matrix}\overline{\gamma_{sr}}=\overline{\gamma_{rd}}\\ \overline{\gamma_{sr}}=\overline{\gamma_{re}}\end{matrix} \\ \frac{\overline{\gamma_{sr}}}{\ln 2}\left[\frac{J\left(\frac{1}{\overline{\gamma_{rd}}}\right)-J\left(\frac{1}{\overline{\gamma_{sr}}}\right)}{\overline{\gamma_{sr}}-\overline{\gamma_{rd}}}-\frac{J\left(\frac{1}{\overline{\gamma_{re}}}\right)-J\left(\frac{1}{\overline{\gamma_{sr}}}\right)}{\overline{\gamma_{sr}}-\overline{\gamma_{re}}}\right], \text{if } \begin{matrix}\overline{\gamma_{sr}}\neq\overline{\gamma_{rd}}\\ \overline{\gamma_{sr}}\neq\overline{\gamma_{re}}\end{matrix} \\ \frac{1}{\ln 2}\left[1-\frac{1}{\overline{\gamma_{re}}}J\left(\frac{1}{\overline{\gamma_{re}}}\right)+\frac{\overline{\gamma_{rd}}J\left(\frac{1}{\overline{\gamma_{rd}}}\right)-\overline{\gamma_{sr}}J\left(\frac{1}{\overline{\gamma_{sr}}}\right)}{\overline{\gamma_{sr}}-\overline{\gamma_{rd}}}\right], \text{if } \begin{matrix}\overline{\gamma_{sr}}\neq\overline{\gamma_{rd}}\\ \overline{\gamma_{sr}}=\overline{\gamma_{re}}\end{matrix} \\ \frac{-1}{\ln 2}\left[1-\frac{1}{\overline{\gamma_{rd}}}J\left(\frac{1}{\overline{\gamma_{rd}}}\right)+\frac{\overline{\gamma_{re}}J\left(\frac{1}{\overline{\gamma_{re}}}\right)-\overline{\gamma_{sr}}J\left(\frac{1}{\overline{\gamma_{sr}}}\right)}{\overline{\gamma_{sr}}-\overline{\gamma_{re}}}\right], \text{if } \begin{matrix}\overline{\gamma_{sr}}=\overline{\gamma_{rd}}\\ \overline{\gamma_{sr}}\neq\overline{\gamma_{re}}\end{matrix} \end{cases} \quad (8)$$

4 Illustrative Results

In this section, numerical results are provided to confirm the accuracy of the proposed calculation. Besides the closed-form expression of the secrecy rate in (8), we also use Monte Carlo simulations as a benchmark to calculate the secrecy rate averaged over all instances of channel gains. It is assumed that the noise variance $N_0 = 1$, and $P_s = P_r = P$. In addition, the secrecy rate curve is drawn versus SNR = P / N_0. We shall consider different values of ϕ_{sr}, ϕ_{rd} and ϕ_{re} of the fading channels that correspond to four different cases given in the formula in (8).

First, consider the trivial scenario where $\phi_{sr} = \phi_{rd} = \phi_{re} = 1$. As a result, $\overline{\gamma_{sr}} = \overline{\gamma_{rd}} = \overline{\gamma_{re}}$. It is obvious from (8) that in this case, the secrecy rate is equal to zero. This result matches very well with the Monte Carlo simulation results, as can be seen from Fig. 2.

Fig. 3, Fig. 4, Fig. 5 shows the secrecy rates' behaviors in the following three cases:

i) ($\overline{\gamma_{sr}} \neq \overline{\gamma_{rd}}$ and $\overline{\gamma_{sr}} \neq \overline{\gamma_{re}}$ with $\phi_{sr} = 1, \phi_{rd} = 3$ and $\phi_{re} = 2$);
ii) ($\overline{\gamma_{sr}} \neq \overline{\gamma_{rd}}$ and $\overline{\gamma_{sr}} = \overline{\gamma_{re}}$ with $\phi_{sr} = \phi_{re} = 1$ and $\phi_{rd} = 2$); and
iii) ($\overline{\gamma_{sr}} = \overline{\gamma_{rd}}$ and $\overline{\gamma_{sr}} \neq \overline{\gamma_{re}}$ with $\phi_{sr} = \phi_{rd} = 2$ and $\phi_{re} = 1$); respectively. In these three cases, the randomization of the fading channels can be exploited to provide a positive secrecy rate. Also, in all three cases, it can be seen that the secrecy rates obtained from Monte Carlo simulation almost coincide with the rates obtained by the expression in (8). This confirms the accuracy of the proposed method.

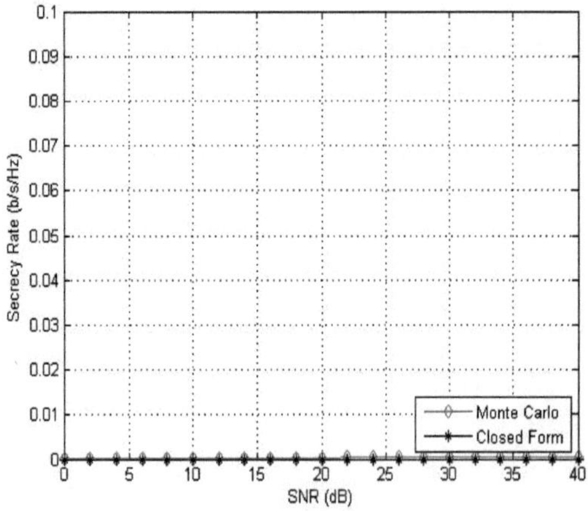

Fig. 2. Secrecy rate versus SNR when $\overline{\gamma_{sr}} = \overline{\gamma_{rd}} = \overline{\gamma_{re}}$

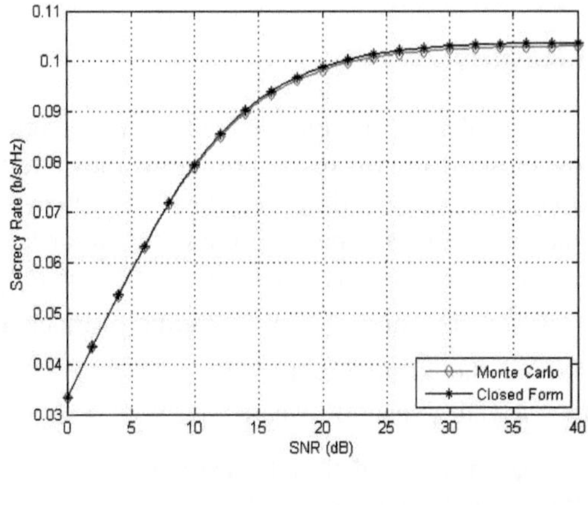

Fig. 3. Secrecy rate versus SNR when $\overline{\gamma_{sr}} \neq \overline{\gamma_{rd}}$ and $\overline{\gamma_{sr}} \neq \overline{\gamma_{re}}$

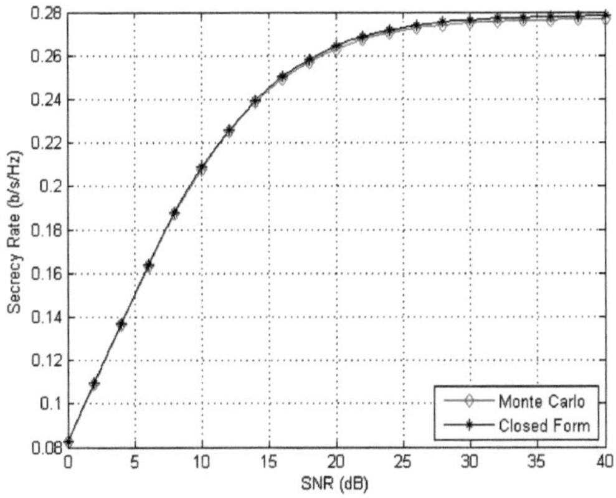

Fig. 4. Secrecy rate versus SNR when $\overline{\gamma_{sr}} \neq \overline{\gamma_{rd}}$ and $\overline{\gamma_{sr}} = \overline{\gamma_{re}}$

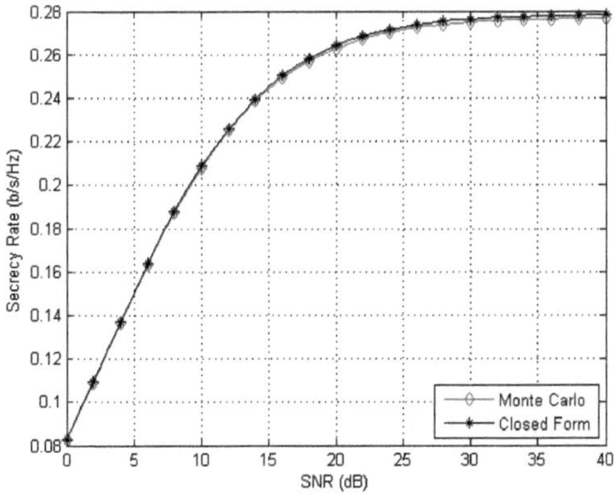

Fig. 5. Secrecy rate versus SNR when $\overline{\gamma_{sr}} = \overline{\gamma_{rd}}$ and $\overline{\gamma_{sr}} \neq \overline{\gamma_{re}}$

5 Conclusion

This paper addressed the secrecy rate of a HD AF relay wire-tap channel in Rayleigh fading. A closed-form formula of the secrecy rate was derived by calculating the expectation of an exponentially distributed random variable using the exponential integral function. Numerical results show that the proposed closed-form formula can be used to effectively calculate the secrecy rate of the considered AF relay channel without resorting to lengthy Monte Carlo simulations. Given the derived closed-form formula, our objective in future works is to find the optimal power allocation scheme that maximizes the secrecy rate of the channel.

References

1. Chen, J., Zhang, R., Song, L., Han, Z., Jiao, B.: Joint relay and jammer selection for secure two-way relay networks. IEEE Trans. Inf. Forensics Security 7(1), 310–320 (2012)
2. Ding, Z., Leung, K., Goeckel, D., Towsley, D.: On the application of cooperative transmission to secrecy communications. IEEE J. Sel. Areas Commun. 30(2), 359–368 (2012)
3. Dong, L., Han, Z., Petropulu, A., Poor, H.: Improving wireless physical layer security via cooperating relays. IEEE Trans. Signal Process. 58(3), 1875–1888 (2010)
4. Hasna, M., Alouini, M.S.: End-to-end performance of transmission systems with relays over Rayleigh-fading channels. IEEE Trans. Wireless Commun. 2(6), 1126–1131 (2003)
5. Krikidis, I., Thompson, J., McLaughlin, S.: Relay selection for secure cooperative networks with jamming. IEEE Trans. Wireless Commun. 8(10), 5003–5011 (2009)
6. Li, J., Petropulu, A., Weber, S.: On cooperative relaying schemes for wireless physical layer security. IEEE Trans. Signal Process. 59(10), 4985–4997 (2011)
7. Loa, K., Wu, C.C., Sheu, S.T., Yuan, Y., Chion, M., Huo, D., Xu, L.: IMT-advanced relay standards [WiMAX/LTE update]. IEEE Commun. Mag. 48(8), 40–48 (2010)
8. Pabst, R., Walke, B.H., Schultz, D., Herhold, P., Yanikomeroglu, H., Mukherjee, S., Viswanathan, H., Lott, M., Zirwas, W., Dohler, M., Aghvami, H., Falconer, D., Fettweis, G.: Relay-based deployment concepts for wireless and mobile broadband radio. IEEE Commun. Mag. 42(9), 80–89 (2004)
9. Rodriguez, L.J., Tran, N.H., Le-Ngoc, T.: Optimal power adaptation for cooperative AF relaying with channel side information. IEEE Trans. Veh. Tech. 62(7), 3164–3174 (2013)
10. Rodriguez, L.J., Tran, N.H., Le-Ngoc, T.: Optimal power allocation and capacity of full-duplex AF relaying under residual self-interference. IEEE Wireless Commun. Lett. 3(2), 233–236 (2014)
11. Rodriguez, L.J., Tran, N.H., Le-Ngoc, T.: Optimal power allocation schemes for single-relay AF wire-tap channels. In: Proc. IEEE Int. Conf. Commun. Electron., Da Nang, Vietnam, July 30-August 1 (2014)
12. Rodriguez, L.J., Tran, N.H., Le-Ngoc, T.: Performance of full-duplex AF relaying in the presence of residual self-interference. IEEE J. Sel. Areas Commun. 32(9), 1752–1764 (2014)
13. Shao, J., Alouini, M.S., Goldsmith, A.: Impact of fading correlation and unequal branch gains on the capacity of diversity systems. In: Proc. IEEE Veh. Technol. Conf., vol. 3, pp. 2159–2163 (July 1999)
14. Simon, M.K., Alouini, M.S.: Digital Communication over Fading Channels: A unified spproach to performance analysis, 1st edn. Wiley, New York (2000)
15. Tekin, E., Yener, A.: The general Gaussian multiple-access and two-way wiretap channels: Achievable rates and cooperative jamming. IEEE Trans. Inf. Theory 54(6), 2735–2751 (2008)
16. Tran, N.H., Rodriguez, L.J., Le-Ngoc, T.: Precoding and symbol grouping for NAF relaying in BICM systems. IEEE Trans. Veh. Tech. 62(6), 2607–2617 (2013)
17. Wang, H.M., Yin, Q., Xia, X.G.: Distributed beamforming for physical-layer security of two-way relay networks. IEEE Trans. Signal Process. 60(7), 3532–3545 (2012)
18. Wyner, A.: The wire-tap channel. Bell Syst. Tech. J. 54(8), 1355–1387 (1975)
19. Zheng, G., Choo, L.C., Wong, K.K.: Optimal cooperative jamming to enhance physical layer security using relays. IEEE Trans. Signal Process. 59(3), 1317–1322 (2011)

A Robust On-Demand Routing Protocol for Cognitive Radio Ad Hoc Networks

Zamree Che-aron[1,*], Aisha Hassan Abdalla[2], Khaizuran Abdullah[3], Wan Haslina Hassan[4], and Md. Arafatur Rahman[5]

[1,2,3] Department of Electrical and Computer Engineering, International Islamic University, Malaysia (IIUM), Kuala Lumpur 53100, Malaysia
[4] Malaysia-Japan International Institute of Technology (MJIIT), Universiti Teknologi Malaysia (UTM), Jalan Semarak, 54100 Kuala Lumpur, Malaysia
[5] Department of Biomedical Electronics and Telecommunications Engineering, University of Naples Federico II, Naples 80138, Italy
one_zamree@hotmail.com, {aisha,khaizuran}@iium.edu.my, wanhaslina@ic.utm.my, arafatur.rahman@unina.it

Abstract. Cognitive Radio (CR) technology has been introduced to solve the problems of spectrum underutilization and spectrum scarcity caused by improper spectrum management policies. In Cognitive Radio Ad Hoc Networks (CRAHNs), which operate without centralized infrastructure support, data routing encounters various challenges including frequent topology changes, heterogeneous spectrum availability, and intermittent connectivity caused by the activities of Primary Users (PUs). In this paper, a robust on-demand routing protocol for CRAHNs, referred to the Robustness Aware Cognitive Ad-hoc Routing Protocol (RACARP), is proposed with an aim to provide robust paths for data delivery. The Expected Path Delay (EPD) routing metric used for path decision is introduced and applied in the protocol. The metric takes account of the link delay and the effect of packet loss on wireless links. Furthermore, the protocol avoids creating a transmission path that uses PU's channel in PU regions in order to counteract the impact of PU activities which can simply cause communication interruptions. The protocol also jointly exploits path and spectrum diversity in routing process in order to provide multi-path and multi-channel routes for the purpose of fast route recovery. The performance evaluation is conducted through simulation using NS-2 simulator. The simulation results prove that the RACARP protocol achieves better performance in terms of average throughput and average end-to-end delay as compared to the Dual Diversity Cognitive Ad-hoc Routing Protocol (D2CARP).

Keywords: robustness, cognitive radio ad hoc network, routing protocol, expected path delay, route recovery, PU impact avoidance.

[*] Corresponding author.

K.J. Kim and N. Wattanapongsakorn (eds.), *Mobile and Wireless Technology 2015*, Lecture Notes in Electrical Engineering 310, DOI: 10.1007/978-3-662-47669-7_4

1 Introduction

Due to huge advancement of wireless technologies and rapidly increasing demand of wireless services, the radio spectrum has become an expensive and scarce resource. Moreover, the static spectrum allocation policy which allows specific wireless services to access fixed spectrum bands for data transmission leads to the spectrum underutilization problem [1]. Cognitive Radio (CR) technology [2-4] has been proposed as a promising solution to improve the spectrum utilization, reduce the congestion in the unlicensed bands and alleviate the shortage of spectrum resources. The main concept of Cognitive Radio Ad Hoc Network (CRAHN) [5] is that, in a wireless ad hoc network, the unlicensed users (or Secondary Users (SUs)) are allowed to access the temporally unused licensed spectrum bands for data communications without harmful interference to the licensed users (or Primary Users (PUs)). With unique characteristics of CRAHNs, the traditional ad hoc routing protocols are not suitable to apply in the networks and new several challenges [6-7] must be taken into account. The main challenge is to deal with the dynamic spectrum availability. In CRAHNs, the Spectrum Opportunities (SOPs) [8] (a set of spectrum bands currently unoccupied by PUs) are time and location varying due to dynamic PU activities. Consequently, the collaboration between spectrum decision and path selection is needed. Another challenge is how to determine the optimal path for data transmission, which provides high network performance. Therefore, the effective routing metrics able to accurately account for the quality of different paths are required. Moreover, in CRAHNs, a link failure frequently occurs caused by not only the node mobility but also the appearance of PU activity. The data transmission of SUs will be interrupted immediately after a PU activity is detected in order not to cause harmful interference to the PU. As a result, the efficient path recovery mechanism is needed in order to rapidly recover the failed paths.

This paper proposes a robust on-demand routing protocol for CRAHNs, namely the Robustness Aware Cognitive Ad-hoc Routing Protocol (RACARP), which is an extension of AODV protocol [9]. In RACARP, the new routing metric called the Expected Path Delay (EPD) metric, which takes account of the effect of packet loss and link delay, is proposed and applied. Furthermore, the protocol establishes a transmission path that avoids using the PU's channel in the PU regions in order to counteract the impact of PU activity which can simply cause communication interruptions. The proposed path-selection mechanism favors a transmission path with small delay, low packet loss and less service interruption caused by PU activity. In addition, the protocol exploits the joint path and spectrum diversity in routing so that the source node is able to dynamically switch to different paths and channels in presence of path failure during data transmission for the purpose of fast route recovery. The simulation results obviously show that the RACARP protocol outperforms the Dual Diversity Cognitive Ad-hoc Routing Protocol (D2CARP) [10] in terms of average throughput and average end-to-end delay.

The remainder of the paper is organized as follows. The overview of RACARP protocol is provided in Section 2. Section 3 describes the robust routing techniques in the protocol. Then, in Section 4, the protocol operations are explained in detail. Section 5 presents the simulation configuration, simulation parameters, and protocol performance evaluation followed by the conclusion in section 6.

2 Robustness Aware Cognitive Ad-Hoc Routing Protocol

The RACARP protocol is an on-demand routing protocol, which triggers the route discovery process only when a data transfer is required by a source node, and also shares some common functionalities with D2CARP protocol [10]. Furthermore, the protocol exploits the joint path and spectrum diversity in routing to provide multi-channel multiple paths based on the EPD routing metric and also avoids creating the transmission paths that uses the PU's channel in the PU regions during route formation with an aim to improve the network performance, provide fast route recovery in presence of PU activity, and make the transmission paths less vulnerable to the impact of PUs. A path with lowest EPD value is selected for data transmission. The sequence number, which indicates the freshness of route information, is utilized to circumvent the problem of routing loops. As different from the D2CARP protocol, the RACARP adds three more control packets, i.e. ETX (Expected Transmission Count) probe packet, RTT (Round-Trip Time) probe packet, and RTT acknowledgement packet. Therefore, the RACARP's main control messages include RREQ (Route REQuest), RREP (Route REPly), RERR (Route ERRor), ETX probe, RTT probe and RTT acknowledgement packet.

By applying RACARP protocol, each node in the network does not need to know the complete path from source to destination for data transmission. Instead, it only utilizes the local routing information (e.g. next-hop node and forward channel) stored in its routing table.

3 Robust Routing Techniques

This section shall describe how the mechanism of path establishment is optimized with an aim to provide robust communications in CRAHNs.

3.1 Path Decision Based on EPD Routing Metric

The EPD routing metric used in the RACARP protocol takes account of the link delay and the effect of packet loss on wireless links during route setup stage. The EPD(p) represents the expected time it takes a probe packet to travel along a path p from a node to another node which can be defined as:

$$\mathrm{EPD}(p) = \sum_{\mathrm{Link}\ l \in p} \mathrm{ELD}(l) \tag{1}$$

where p is a path which are composed of the set of links. The ELD(l) that denotes the Expected Link Delay of the link l can be calculated as:

$$\mathrm{ELD}(l) = \mathrm{ETX}(l) * \left(\frac{\mathrm{RTT}(l)}{2} \right) \tag{2}$$

where RTT(l) is the interval between the sending of an RTT probe packet and the receipt of the corresponding RTT acknowledgement packet over the link l. The ETX(l) is the Expected Transmission Count of the link l, representing the expected number of retransmissions required to successfully transmit an ETX probe packet over the link l. The ETX(l) can be measured as:

$$\mathrm{ETX}(l) = \frac{1}{[1 - P_f(l)][1 - P_r(l)]} = \frac{1}{[d_f(l) * d_r(l)]} \tag{3}$$

where $P_f(l)$ and $P_r(l)$ are the probability of packet loss in the forward and reverse direction of the link l respectively. The $d_f(l)$ (forward delivery ratio) denotes the probability that an ETX probe packet is successfully transmitted to the neighbor over the link l during the window period (i.e. *ETX_PROBE_WINDOW*). The $d_r(l)$ (reverse delivery ratio) represents the probability that an ETX probe packet sent from the neighbor is successfully received through the link l during the window period (i.e. *ETX_PROBE_WINDOW*). The $d_f(l)$ and $d_r(l)$ can be calculated as:

$$d_f(l) = \frac{n_f(l)}{\mathrm{ETX\_PROBE\_WINDOW}} \tag{4}$$

$$d_r(l) = \frac{n_r(l)}{\mathrm{ETX\_PROBE\_WINDOW}} \tag{5}$$

where $n_f(l)$ is the number of ETX probe packets successfully sent to the neighbor through the link l. While $n_r(l)$ represents the number of ETX probe packets successfully received from the neighbor via the link l.

To measure the updated ETX and RTT value of a link, each SU periodically broadcasts an ETX and RTT probe packet to the neighbors according to the ETX probe interval and RTT probe interval respectively.

In the simple network shown in Fig. 1, Node A (source node) requires a transmission path towards Node I (destination node). It is under the assumption that only one channel is available for communications and the ELD value of each link is exhibited. For the RACARP protocol, the path A-D-G-I with lowest EPD value is chosen for data transmission. Therefore, the protocol always selects a transmission path with highest quality in terms of link delay and packet loss probability.

3.2 PU Channel Avoidance in PU Regions

The RACARP protocol always avoids establishing a transmission path that uses the PU's channel in the PU regions in order to alleviate the impact of PU activities, which can result in frequent communication interruptions. To accomplish this goal, in every specific time interval, each SU checks to determine whether or not it is currently located in a PU's transmission range. In case a SU is aware that it exists in a PU region, the PU's channel will be stored in its list of blocked channels; otherwise, it removes the PU's channel from the blocked channel list. In RACARP, a SU inside a PU region always declines to create a transmission path that uses the PU's channel. Therefore, the data packets are not delivered through the PU's channel when crossing the PU region.

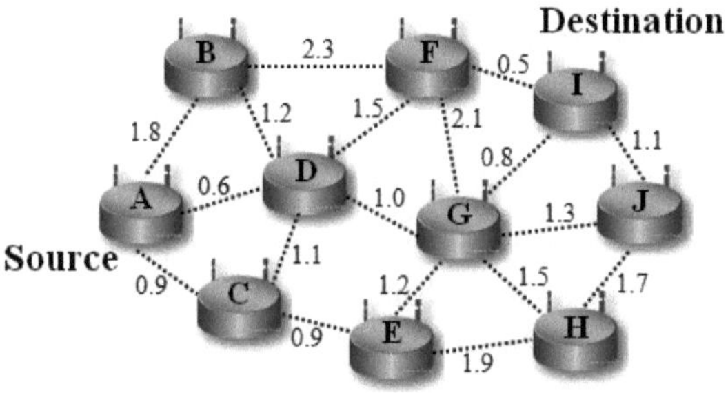

Fig. 1. Path decision based on EPD routing metric

As exhibited in Fig. 2(a), only one licensed channel is available in the network. By avoiding using PU's channel in PU regions, the RACARP protocol establishes the transmission path as represented as the solid line from S to D in order to reduce service interruptions caused by PU activities.

In case both licensed and unlicensed channels are available in the network, to create a transmission path by avoiding PU territories may produce a large end-to-end delay due to the establishment of long detour. In such a case, the RACARP protocol may establish a transmission path which crosses the PU regions. However, the PU's channel is not used to deliver the data packets when crossing the PU regions (as shown in Fig. 2(b)).

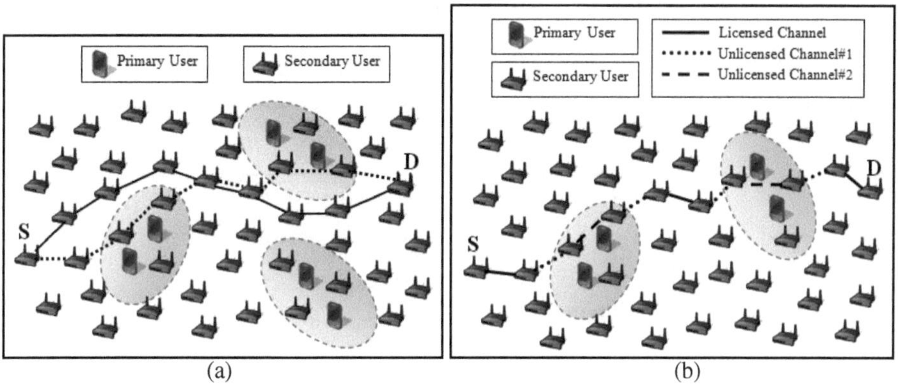

Fig. 2. Path establishment based on PU channel avoidance in PU regions: a) Only one licensed channel is available, b) Both licensed and unlicensed channels are available

3.3 Exploitation of Joint Path and Spectrum Diversity

The RACARP protocol utilizes the joint path and spectrum diversity in routing process in order to provide multi-path and multi-channel routes. As a result, the source node is able to immediately switch among different paths and different channels in appearance of path failure during data transmission in order for fast route recovery. Therefore, the performance degradation caused by the activity of PUs can be alleviated.

Fig. 3(a), (b) and (c) depict the advantage of exploiting the joint path and spectrum diversity in the RACARP protocol. As shown in Fig. 3(a), after the route discovery process is successfully completed, SU_1 (source node) begins transmitting data packets along the optimal path with minimum EPD value (i.e. $SU_1 \xrightarrow{ch1} SU_2 \xrightarrow{ch2} SU_4$) towards SU_4 (destination node). During data delivery, in case SU_2 is moved into the PU_1-PU_2 region and the PU activity on channel 1 is detected (see Fig. 3(b)), it notifies its neighbors (i.e. SU_1 and SU_4) of the PU activity detection. Then, SU_1 immediately uses another available channel (i.e. channel 2) to transmit data packets without changing path direction. Consequently, the new transmission path is $SU_1 \xrightarrow{ch2} SU_2 \xrightarrow{ch2} SU_4$. Afterwards, if SU_2 detects another PU activity over channel 2 (see Fig. 3(c)), it must instantaneously disable the channel 2 for data transmission and notify its neighbors of the PU activity detection. Subsequently, since SU_2 is unable to operate over both channel 1 and channel 2, SU_1 immediately switches to another available path (i.e. $SU_1 \xrightarrow{ch2} SU_3 \xrightarrow{ch1} SU_4$) for data delivery without triggering a new route discovery process.

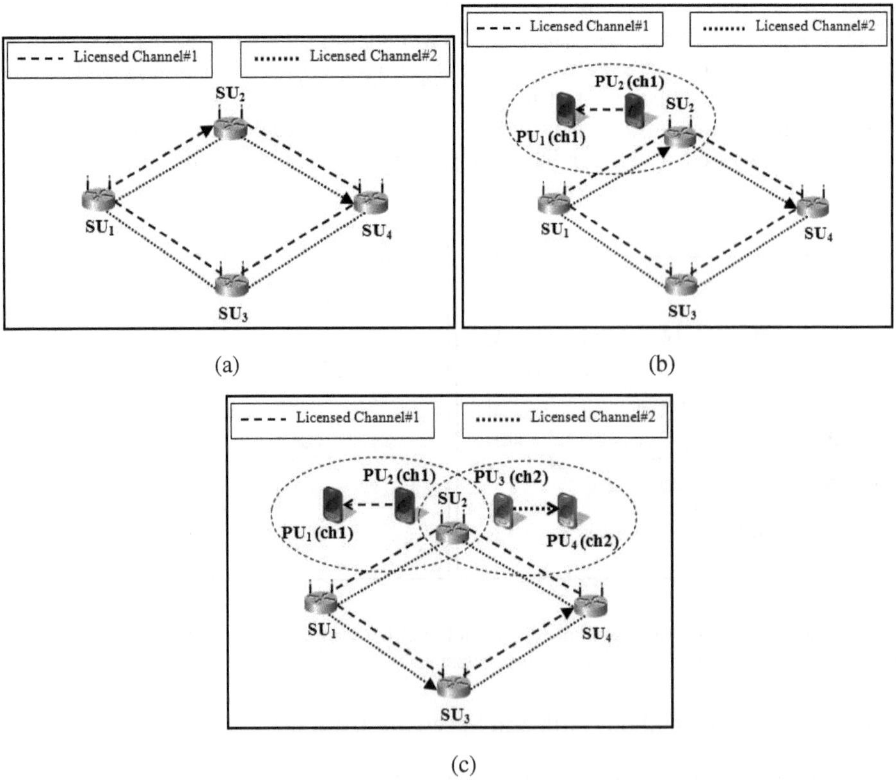

Fig. 3. Advantage of exploiting the joint path and spectrum diversity: a) No PU activity is detected, b) PU activity over channel 1 is detected, c) PU activities over channel 1 and channel 2 are detected

4 Protocol Operations

The key operations of the RACARP protocol include the route discovery, route maintenance and route recovery mechanism which are described below in further detail. Moreover, in the network model, it is assumed that each SU is equipped with multiple wireless interfaces. Each interface can only operate over one of non-overlapping channels.

4.1 Route Discovery

When a source node needs to transmit a data packet towards another node for which the routing information is unknown, it broadcasts an N-RREQ packet for the destination node to its neighbors through all its available channels (i.e. not used by a PU). An intermediate node which receives the first N-RREQ packet creates a routing table entry for a route towards the source node, called a reverse route, and records the channel, through which the packet has been transmitted, in its routing table. Then, it re-broadcasts the packet via all its vacant channels (i.e. free from a PU). If an extra N-RREQ packet with the same sequence number received from the same node but on different channel, it creates another routing table entry of reverse route for that channel without rebroadcasting the packet. In such a way, the multi-channel reverse routes are established. The record of reverse route will be updated only if it receives an additional N-RREQ packet with a higher sequence number or the same sequence number but lower EPD value. The stale N-RREQ packet received by a node will be discarded to avoid the problem of routing loops.

The N-RREQ packet is re-broadcasted until it reaches the destination node. Alternatively, it may arrive at an intermediate node that has a record of a valid route towards the destination. In both cases, an N-RREP packet is generated and sent back to the previous node via the same channel that the N-RREQ packet has been received. Also, a further N-RREQ packet received from the same node but on different channels will not be ignored. However, if the destination node receives an extra N-RREQ packet from a different node and all the following conditions are satisfied: initially, *RREQ's First-Hop Node ID* in the N-RREQ packet is different from *First-Hop Node ID* of reverse route entries in its routing table. Second, ID of the node from which the N-RREQ packet has been received is different from *Next-Hop Node's ID* of reverse route entries in its routing table. Third, *EPD* value in the N-RREQ packet is less than or equal to the *minimum EPD* value of previously established reverse route entries in its routing table. Then, it will create another routing table entry of reverse route for the channel through which the N-RREQ packet has been sent. As a result, the multi-path reverse routes are created.

In the route reply phase, an intermediate node which receives the first N-RREP packet creates a routing table entry for a route towards the destination node, referred to as a forward route, and records the channel, through which the N-RREP packet has been transmitted, in its routing table. Subsequently, it forwards the copies of the N-RREP packet back towards the source node through all its active reverse routes over available channels (i.e. not occupied by a PU). If an intermediate node receives an extra N-RREP packet from the same sender but on different channel, it creates another routing table entry of forward route for that channel. Then, it re-forwards the N-RREP packet towards the source node only through its reverse route over the same channel. In this fashion, the multi-channel forward routes are built. Only in case a node receives a fresher or better N-RREP packet, which has a greater sequence number or the same sequence number with smaller EPD value, the forward route entry will be updated.

When an N-RREP packet arrives at the source node, a path from source to destination is created and the node can begin sending data packets. To have multi-channel forward routes, the source node will not discard the additional N-RREP packets received from the same sender but on different channels. Moreover, the multi-path forward routes can be also established in case it receives an extra N-RREP packet from a different sender under the following conditions: initially, *RREP's First-Hop Node ID* in the N-RREP packet is different from *First-Hop Node ID* of forward route entries in its routing table. Second, ID of the node from which the N-RREP packet has been received is different from *Next-Hop Node's ID* of forward route entries in its routing table. Third, *EPD* value in the N-RREP packet is less than or equal to the *minimum EPD* value of previously established forward route entries in its routing table.

In addition, to avoid creating a transmission path using PU's channel in PU regions with an aim to alleviate the impact of PU activities, a node inside a PU region always discards an N-RREQ and N-RREP packet received through the PU's channel which is stored in its blocked channel list.

4.2 Route Maintenance and Recovery

As data packets flow from source to destination, each node over the transmission path updates the lifetime (i.e. expiration time) of its forward and reverse routes in order to maintain the link connectivity. However, if a route's lifetime is expired, the routing entry for that route will be invalid.

In the RACARP protocol, each SU sets a timer for every specific time to sense a PU activity. During data transmission, if a SU detects a PU activity, it is unable to transmit a data packet through the channel which overlaps with the PU's transmission frequency, thus resulting in a link failure. For that case, the node responds to the failure by immediately selecting another available channel or path for data delivery without triggering a new route discovery process. However, the previously disabled routing entries can be enabled again after the PU activity is stopped or finished.

In addition, a link failure, which is detected by link-layer feedback, can result from not only PU activity but also node mobility, node fault, link degradation, etc. During data delivery, if a node detects a link breakage and no alternative available channel or path is found, then it marks the broken route in its routing table as invalid. It also generates an N-RERR packet and broadcasts it to all its neighbors. As the N-RERR packet propagates towards the source node, a node that receives the N-RERR packet invalidates all affected routing table entries. When the N-RERR packet arrives at the source node, a new route discovery process will be triggered.

5 Simulation Configuration and Performance Evaluation

The performance evaluation was conducted through simulation using the NS-2 simulator [11] with an extension to support the CR environments. In the simulation, the 100 SUs are placed in a 1000 x 1000 m^2 terrain. The distance between two SUs along X- and Y-axis is set to 100 m. There are 10 PUs randomly located in the simulation area. The PU activities are modeled according to the ON/OFF process [12] with exponential distribution with parameter λ of 75, referred to as PU activity parameter. The ON state denotes the period where the channel is occupied by PU and the OFF state represents the period where the channel is available for SUs' communications. The transmission range of SU and PU is set to 150 m. The traffic load is modeled as CBR (Constant Bit Rate) data packets with size of 512 bytes at the packet interval of 50 ms over UDP (User Datagram Protocol) connections. The duration of simulation run is 150 seconds. There are 4 non-overlapping channels given for multi-channel data communications. The two-ray ground reflection model is specified as the radio propagation type and the IEEE 802.11 is used for MAC protocol. Additionally, to calculate the updated ELD value of each link, an ETX and RTT probe packet are periodically broadcasted to the neighbors every 1 and 0.5 second respectively. The ETX_PROBE_WINDOW period is set to 10 seconds.

The protocol performance is evaluated through simulation (with varying number of data traffic connections) based on the performance metrics including average throughput and average end-to-end delay. The NS2 Visual Trace Analyzer [13] is used to analyze the simulation results. To validate the performance improvement of RACARP protocol, the simulation results are benchmarked with that of D2CARP protocol [10] under identical scenarios.

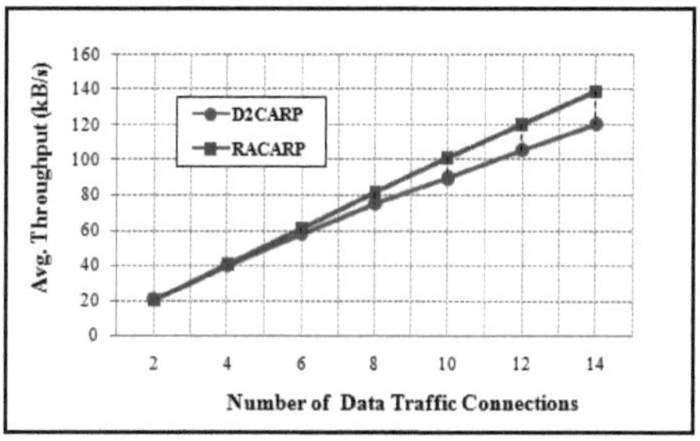

Fig. 4. Simulation results of average throughput versus the number of data traffic connections

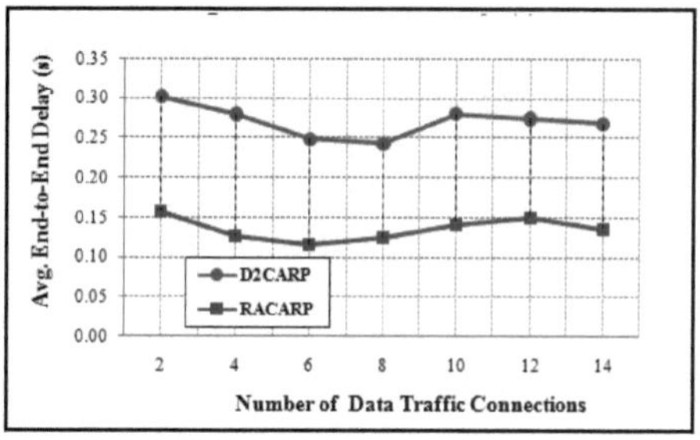

Fig. 5. Simulation results of average end-to-end delay versus the number of data traffic connections

Fig. 4 shows the results of average throughput versus the different number of data traffic connections. The average throughput is defined as the ratio of the total amount of data successfully received by the destination to the time it takes from the data start time to the data stop time. From the figure, when the number of data traffic connections increases, the throughput results of both protocols also rise. However, RACARP protocol outperforms in all cases compared to D2CARP protocol.

The RACARP takes account of the impact of packet losses and link delay for path selection as well as circumventing creating a path that uses PU's channel in PU regions for data delivery in order to alleviate the performance degradation caused by PU activities. Therefore, its transmission path is more robust than the one established by the D2CARP, thus leading to higher throughput results. In the network with 14 data traffic connections, the RACARP achieves a throughput enhancement of about 15.19% over the D2CARP protocol.

The results of average end-to-end delay versus the number of data traffic connections are presented in Fig. 5. The average end-to-end delay is defined as the average time taken by data packets to be delivered across a network from source to destination. From the figure, it is observed that RACARP protocol enhances the average end-to-end delay more than D2CARP protocol. As expected, the RACARP utilizes the EPD routing metric. The EPD routing metric considers the link delay and the effect of packet loss for path decision, i.e. a path with lowest EPD value is chosen for data transmission. Therefore, a transmission path with highest quality in terms of link delay and packet loss probability is always selected. On the contrary, the minimum hop count which is used as a routing metric in the D2CARP is unable to reflect the accurate quality of transmission path, especially in terms of latency, hence causing higher end-to-end delay results. In the network with 14 data traffic connections, the RACARP achieves an end-to-end delay enhancement of about 48.94% over the D2CARP protocol.

6 Conclusion

Cognitive Radio (CR) technology has been proposed as an effective solution for efficient utilization of radio spectrum. Routing in CRAHNs is very challenging due to node mobility, dynamic spectrum availability, energy constraint, no centralized network infrastructure support, etc.

In this paper, the RACARP protocol has been proposed. It is a robustness aware on-demand routing protocol for CRAHNs. Also, the EPD routing metric used for path decision in the RACARP has been introduced. The metric takes account of the link delay and the effect of packet loss on wireless links. In addition, the protocol circumvents creating a path that uses PU's channel in PU regions for data transmission in order to alleviate the impact of PU activities that can cause frequent communication interruptions. Moreover, the protocol exploits the joint path and spectrum diversity in routing process to provide the source node with multi-path multi-channel routes. This is with an aim to offer fast route recovery in appearance of path failure during data delivery. The protocol performance has been evaluated through simulation using NS-2 simulator. The performance comparison between the RACARP and D2CARP protocol has been carried out. The simulation results have confirmed that the RACARP outperforms the D2CARP by achieving higher average throughput and providing lower average end-to-end delay in identical conditions.

References

1. McHenry, M.: Spectrum White Space Measurements. Presentation to New America Foundation's Broadband Forum (2003), http://www.newamerica.net/files/nafmigration/archive/Doc_File_185_1.pdf
2. Haykin, S.: Cognitive Radio: Brain-Empowered Wireless Communications. IEEE Journal on Selected Areas in Communications 23(2), 201–220 (2005)
3. Akyildiz, I.F., Lee, W.-Y., Vuran, M.C., Mohanty, S.: Next Generation/Dynamic Spectrum Access/Cognitive Radio Wireless Networks: A Survey. Computer Networks 50(13), 2127–2159 (2006)
4. Mitola, J.: Cognitive Radio Architecture Evolution. Proceedings of the IEEE 97(4), 626–641 (2009)
5. Akyildiz, I.F., Lee, W.-Y., Chowdhury, K.R.: CRAHNs: Cognitive Radio Ad Hoc Networks. Ad Hoc Networks 7(5), 810–836 (2009)
6. Cesana, M., Cuomo, F., Ekici, E.: Routing in Cognitive Radio Networks: Challenges and Solutions. Ad Hoc Networks 9(3), 228–248 (2011)
7. Sengupta, S., Subbalakshmi, K.P.: Open Research Issues in Multi-hop Cognitive Radio Networks. IEEE Communications Magazine 51(4), 168–176 (2013)
8. Akyildiz, I.F., Lee, W.-Y., Vuran, M.C., Mohanty, S.: A Survey on Spectrum Management in Cognitive Radio Networks. IEEE Communications Magazine 46(4), 40–48 (2008)
9. Perkins, C.E., Belding-Royer, E.M., Das, S.R.: Ad Hoc On-Demand Distance Vector (AODV) Routing. RFC 3561, Internet Engineering Task Force (IETF) (2003), http://www.ietf.org/rfc/rfc3561.txt
10. Rahman, M.A., Caleffi, M., Paura, L.: Joint Path and Spectrum Diversity in Cognitive Radio Ad-Hoc Networks. EURASIP Journal on Wireless Communications and Networking 2012(1), 1–9 (2012)
11. Issariyakul, T., Hossain, E.: Introduction to Network Simulator 2 (NS2). In: Issariyakul, T., Hossain, E. (eds.) Introduction to Network Simulator NS2, ch. 2, pp. 1–18. Springer, USA (2009)
12. Chowdhury, K.R., Felice, M.D.: SEARCH: A Routing Protocol for Mobile Cognitive Radio Ad-Hoc Networks. Computer Communications 32(18), 1983–1997 (2009)
13. Rocha, F.: NS2 Visual Trace Analyzer (2012), http://nsvisualtraceanalyzer.wordpress.com

Guiding Users to Shops Using the Near-Field Communication between Signages and Mobile Terminals

Yoshio Suga[1,*], Daiki Takahashi[2], and Kazumasa Takami[2]

[1] Graduate School of Engineering and Information Systems, Soka University
[2] Faculty of Engineering, Soka University
1-236 Tangi-cho, Hachioji-shi, Tokyo, 192-8577 Japan
e14m5216@soka-u.jp, k_takami@soaka.ac.jp

Abstract. This paper proposes two algorithms for navigating users to their destinations using near-field communication between their mobile terminals and digital signage units in an underground area where GPS is not usable. One algorithm focuses on the angle to the destination while the other looks at the distance to the destination. The paper also proposes an algorithm for avoiding traffic congestion by limiting the number of mobile terminals that communicate with a signage unit simultaneously. We have built an underground mall model on a simulator, and evaluated the algorithms in terms of the rate of successful arrival at the destination and the time it takes to reach the destination.

1 Introduction

As the use of the Internet has become widespread, it has become commonplace for the user of a smart phone equipped with a highly precise GPS system to navigate his/her way using a map application. Some map applications simply display an arrow to indicate the direction toward the destination [1]. A problem with this approach is that it cannot be used in areas where radio waves from communication carriers cannot be received and consequently GPS is not usable, such as underground malls.

There are several studies on determining the location of a mobile terminal in an underground mall. One method is to use the degree of intensity of radio waves received from Wi-Fi access points (APs), and some of such methods have been implemented commercially [2-4]. Another method is to apply the centroid method to location data received from multiple mobile terminals using near-field communication [5]. However, this method requires the user to check the screen of his/her smartphone while walking, which is a dangerous action that is spreading and causing anxiety in society. Navigation of users to their destinations has not been studied in connection with this method.

Since electrical advertising media, such as digital signages (hereafter simply referred to as "signages") have been widely implemented, a study has been made to guide users to their destination shops using intermittent near-field communication between their mobile terminals and signages [6]. However, this study assumes that

* Corresponding author.

K.J. Kim and N. Wattanapongsakorn (eds.), *Mobile and Wireless Technology 2015*,
Lecture Notes in Electrical Engineering 310, DOI: 10.1007/978-3-662-47669-7_5

only one user accesses signage units. Therefore, the signage server is not designed to manage connection resources in ways that allow multiple users to access signages.

This paper proposes two algorithms for navigating users to their destinations using intermittent near-field communication between their mobile terminals and signage units in an underground mall, where GPS is not usable. One focuses on the angle to the destination while the other looks at the distance to the destination. The paper also proposes an algorithm for avoiding traffic congestion by limiting the number of mobile terminals that communicate with a signage unit simultaneously. We have built an underground mall model on a simulator, and evaluated the algorithms in terms of the rate of successful arrival at the destination (arrival rate) and the time it takes to reach the destination (arrival time).

This paper is structured as follows. Section 2 describes the proposed navigation algorithms, focusing on human-to-machine (H2M) and machine-to-machine (M2M) interfaces. Section 3 presents the evaluation system developed, which uses virtual mobile terminals. This section also proposes an underground shopping mall model. Section 4 evaluates the proposed algorithms in terms of the arrival rate and arrival time using simulation. Finally, Section 5 presents the conclusions.

2 Navigation Using Intermittent Near-Field Communication between Mobile Terminals and Signage

2.1 Overview of the Proposed Navigation Services

When a user enters the underground mall concerned, his/her mobile terminal accesses the nearest signage. A list of shops is displayed on the terminal. A user selects a shop user wants to visit, and sends that information to the signage. Signages communicate with the mobile terminal to guide the user to the destination shop. They display the appropriate direction with one of the four possible arrows (ahead, back, left or right) so that the user need not look at the screen of mobile terminal continuously [6].

When a user enters the underground mall, mobile terminal sends data that contains user's preference information to the nearest signage. Each signage analyzes preference information in data received from users in its immediate surroundings, selects an advertisement that it judges to be the most effective for the users, and displays it. It also displays, on the mobile terminals of individual users, a number of shop advertisements that it judges to be effective for each individual. Each user selects a shop user wants to visit from the displayed advertisements, and sends that information to the nearest signage. The signage assigns a user ID to the user. After that, each time a user comes close to a signage, the unit displays user's userID and an arrow user should follow to go to user's destination. Since the screen area of a signage unit is limited, the ID and the associated arrow are displayed only for a certain period, T_D.

2.2 System Architecture Based on M2M Communication Model

Two functions are required for implementing the proposed service in an M2M model [7]: the M2M function that enables autonomous near-field communication between the mobile terminal and the nearest signage unit, and the H2M function that provides user interfaces. A system architecture showing how these functions are allocated is shown in Fig. 1.

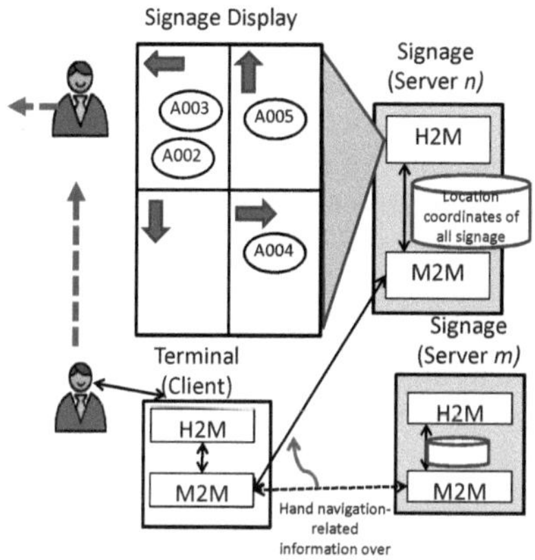

Fig. 1. System architecture based on an M2M model

Signages are installed at intersections near shops. When a mobile terminal comes into an area covered by the near-field communication of a signage unit, it sends to the signage unit a message that contains the ID of the user (UserID), the ID of user's destination signage (SignageID), and information about the area for which user is heading (Area). The signage that has received this message compares Area in the message and the area in which it is located. If these are identical, the signage calculates the area to which the user should move (nextArea) using one of the navigation algorithms described in Section 2.4, and sends nextArea to user's mobile terminal. This nextArea is put in Area in a message that the mobile terminal will send to the signage it will encounter next. This information is handed over to the next signage along with UserID as navigation-related information. If the area for which the user is heading is different from the information in Area, no navigation is provided. The formats of messages sent from a signage to a mobile terminal and messages sent from a mobile terminal to a signage to achieve the handover of navigation-related information are shown in Fig. 2.

Type Signage 1byte	nextArea 2byte	UserID* 2byte

*Only at the time of registration

(a) Message sent from a signage to a mobile terminal

Type Client 1byte	Area 1byte	SignageID 2byte	UserID 2byte

(b) Message sent from a mobile terminal to a signage

(note) *Type*: the source of the message, *nextArea*: area to which the user should go, *UserID*: userID *Area*: area for which the user is heading, *SignageID*: ID of the signage unit near the destination shop

Fig. 2. Message formats

2.3 M2M and H2M Functions

The M2M function provides not only near-field communication but also connection resource management, which controls the number of simultaneous connections to avoid traffic congestion. In an area with a high density of mobile terminals, multiple terminals communicate with a signage simultaneously. To avoid being overloaded with processing of many connection requests, the signage limits the number of mobile terminals that can be connected to it simultaneously to N_L. When a new mobile terminal comes into its area, it accepts it only when the number of mobile terminals connected does not exceed N_L. Specifically, each signage has a table for managing the state of each mobile terminal. It adds mobile terminals communicating with it in the table. When the number of mobile terminals in a connected state exceeds N_L, no new mobiles are added to the table.

The main H2M function is providing navigation. We propose two navigation algorithms: an angle-based algorithm and a distance-based algorithm. We have decided to evaluate these two different algorithms because they are likely to provide different qualities of navigation depending on the positions of signages that stand between the current location and the destination.

(1) Angle-Based Navigation Algorithm

Each signage unit calculates the angle between a line drawn from it to each adjacent signage unit and a line drawn from it to each signage unit, and determines which adjacent signage unit should be chosen when one of the signage units is the destination. This is determined for every signage. This information is stored in a database in advance. When a user comes into the area of a signage unit, it checks the area that the user is heading for and the area that the signage unit is to select for this user. If these are identical, the signage communicates with the user's mobile terminal to provide navigation.

Navigation based on the angle-based algorithm is shown Fig. 3. Let the angle between a lines drawn from the current location, S, to the destination, G, and a line drawn from the current location to an adjacent signage unit, D1, be θ1, and the angle between the former line and a line drawn from the current location to another adjacent signage unit, D2, be θ2. Since θ1<θ2 in Fig. 3, the current signage unit selects D1, and guides the user toward D1.

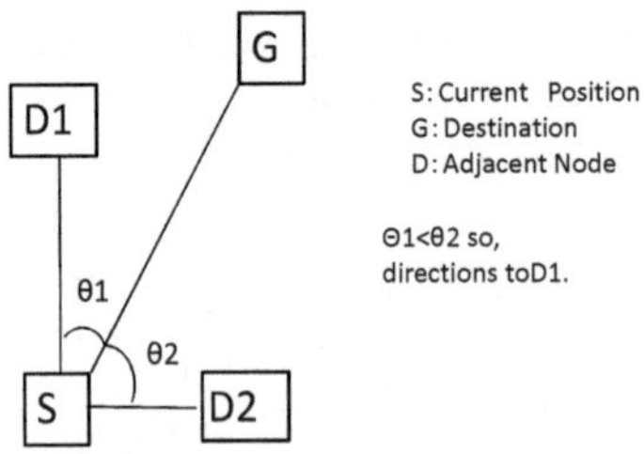

Fig. 3. Navigation based on the for angle-based algorithm

A user is guided toward user's destination as follows.

STEP1: When a mobile terminal comes into an area covered by the near-field communication of a signage, the signage registers ID of the terminal in a table.

STEP 2: The signage unit sets the state of the terminal to a connected state in the table, and at the same time initializes the counter for the terminal, T_D.

STEP 3: The gradient of the straight line between the current and destination locations is calculated by applying the location coordinates of the current signage unit, $S(x_1, y_1)$, and those of the destination,
$G(x_2, y_2)$, to Eq. (1).

$$\theta = \frac{180}{\pi} \tan^{-1}\left(\frac{y_2 - y_1}{x_2 - x_1}\right) \tag{1}$$

STEP 4: The signage then calculates the gradient of the straight line between the current location and each of the adjacent signages located in the four directions from the current location by applying the location coordinates of the adjacent signage unit to Eq. (1).

STEP 5: The signage checks the difference between the gradient calculated in STEP 3 and that calculated in STEP 4, and selects the signage that has the smallest difference as the next signage toward which the user should be guided.

STEP 6: The signage calculates the difference, dx, between the x-coordinates of the current and next signage units and the difference, dy, between the y coordinates of the current and next signage units.
If dx>dy, go to STEP 7. Else if dx<dy, go to STEP 8.

STEP 7: If dx>0, “Turn right” is displayed for T_D seconds.
dx<0, “Turn left” is displayed for T_D seconds.

STEP 8: If dy>0, “Turn back” is displayed for T_D seconds.
dy<0, “Go straight ahead” is displayed for T_D seconds.

STEP 9: The signage unit decreases the value of T_D at a certain interval. When the value of T_D becomes 0, the signage removes the information about the mobile terminal concerned from the table, and the mobile terminal becomes idle.

STEP 10: Until the current signage unit becomes the destination signage repeat from STEP 1.

(2) Distance-Based Navigation Algorithm

Navigation based on the distance-based algorithm is shown Fig. 4. A line is drawn between the current location, S, and the destination, G. Let the distance from D1 to that line be L1, and let the distance from D2 to that line be L2. Since L1>L2 in Fig. 4, the signage unit at S selects D2. It determines the direction toward D2 from the coordinates of S and those of D2, and displays an appropriate arrow on its screen.

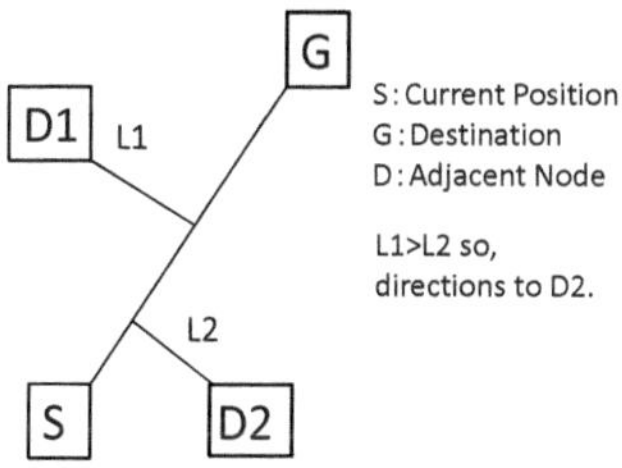

Fig. 4. Navigation based on the distance-based algorithm

A user is guided as described below. STEPs 1 and 2 are omitted below because these are the same as in the previous algorithm.

STEP 3: The signage unit at the current location, $S(x_1, y_1)$, derives an equation for the line between itself and the destination, $G(x_2, y_2)$, using Eq. (2).

$$y = \frac{y_2 - y_1}{x_2 - x_1}x - \frac{y_2 - y_1}{x_2 - x_1}x_1 + y_1 \tag{2}$$

STEP 4: The signage unit at S calculates the distance, L, between each of the adjacent signage units in the four directions from it, $R(x_3, y_3)$, and the S-G line drawn in STEP 1 using Eq. (3).

$$\frac{y_2 - y_1}{x_2 - x_1} = m,\ y_1 - \frac{y_2 - y_1}{x_2 - x_1}x_1 = n, L = \frac{|y_3 - mx_3 - n|}{\sqrt{1 + m^2}} \tag{3}$$

STEP 5: The signage unit at S selects the signage unit that is the closest to the S-G line as the next signage unit toward which the user should be guided.

STEP 6: The steps from here are the same as those in the previous algorithm.

3 Development of an Evaluation System

To evaluate the proposed algorithms, we developed an evaluation system by adding the proposed algorithms to a system that was used to simulate an underground mall [5]. To simulate the M2M communication model and the underground model, we built a simulation base on a general-purpose PC. Virtual Nodes (VNs) based on socket communication can be defined on this base. Mobility of VNs can also be defined on it. We implemented a monitoring function that displays the paths taken by VNs in the underground mall.

In the simulator that we have developed, we mapped the floor plan of Tobu Hope Center [8], which is 105 m long and 70 m wide, on this underground mall model.

An example of simulation results is shown in Fig. 5. The figure shows the path taken by one user who walked from the starting point at S7 to the destination at S30. It indicates that the user was successfully guided from the starting point to the destination.

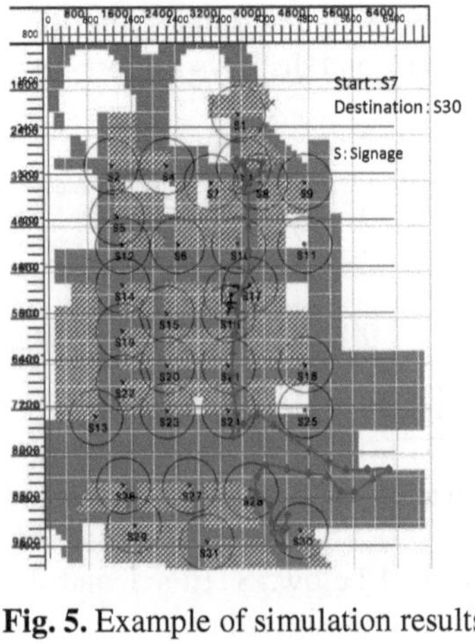

Fig. 5. Example of simulation results

4 Evaluation

The evaluation conditions used are shown in Table 1. There were 31 signage units and 80 mobile terminals. Users walked randomly only on paths.

The processing interval was 1000 msec. Logs were taken every 5 minutes. The walking speed of users was 3.6 km/h. The distance covered by near-field communication was 10 m. Initially, mobile terminals started at random areas. Because there were 80 mobile terminals accessing 31 signage units, it was assumed that, on average, 2 to 3 mobile terminals communicate with a signage unit simultaneously. Since the direction in which a user is guided is indicated by an arrow as is shown in Fig. 2, the maximum number of simultaneous connections, N_L, was determined in such a way that all the arrows are clearly visible on the display.

Table 1. Evaluation conditions

Item	Value	Item	Value
Number of signage	31	Area Size	105m ×70m
Number of users	80	Distance within which communication is possible	10 m
Movement Model	Random walk	Processing interval	1000 milliseconds
Moving speed	3.6km/h	Simulation time	5 minutes
Number of simulations connections	2, 3, or 6	Number of simulation attempts	5
Display duration	30 or 60 seconds		

To evaluate the proposed navigation algorithms, we have defined "arrival rate" and "arrival distance" as follows:

Arrival rate
= (number of mobile terminals that arrived at their destinations successfully)
÷ (number of all mobile terminals)

Arrival time
= Average time it took for mobile terminals to arrive at their destinations

The arrival rate and arrival time of the two algorithms are shown in Figs. 6 and 7 respectively with the duration of an arrow being displayed as a parameter. The maximum number of simultaneous connections, N_L, was varied from 2, 3 to 6.

(1) Arrival Rate
With either algorithm, the larger the maximum number of simultaneous connections, the higher the arrival rate. When the arrow display duration is longer, the arrival rate is lower. This means that, the longer the arrow display duration is, the longer the period between the time when a mobile terminal starts to communicate with a signage and the time when it stops communication. Consequently, new mobile terminals need

to wait for a longer time before they can set up a connection with the nearest signag. This can increase the probability at which a terminal is not navigated, leading to a lower arrival rate. The angle-based algorithm produced a higher arrival rate than the distance-based algorithm. The distance-based algorithm selects the next signage based on the distance between an adjacent signage unit and the line drawn between the signage at the current location and that at the destination location. This means that a signage that exists in the direction opposite to the destination can be selected. If this happens, the mobile terminal concerned is guided to a signage that is farther from the destination than the signage unit at the current location. This mobile terminal is likely not to reach the destination within a certain period. This does not happen with the angle-based algorithm because any adjacent signage that is located in a direction opposite to the destination has a larger angle than any adjacent signage unit that is located in the direction of the destination.

(2) Arrival Time

Neither the arrow display duration nor the maximum number of simultaneous connections has a significant impact on the arrival time. As a whole, the angle-based algorithm produced a shorter arrival time than the distance-based algorithm. It can be concluded that the angle-based algorithm provides more efficient navigation than the distance-based algorithm.

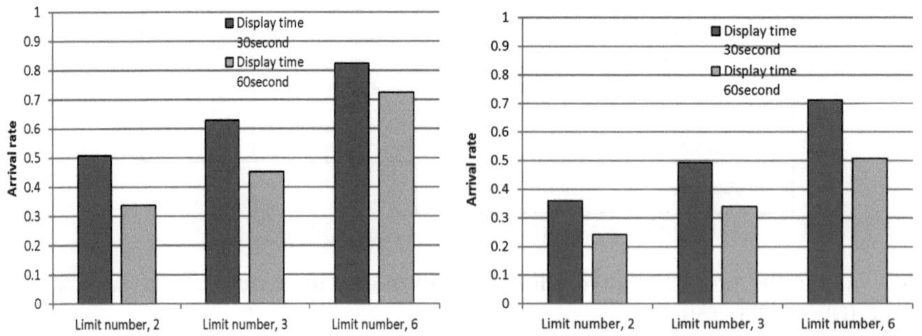

(a) Angle based navigation algorithm (b) Distance based navigation algorithm

Fig. 6. Arrival rate

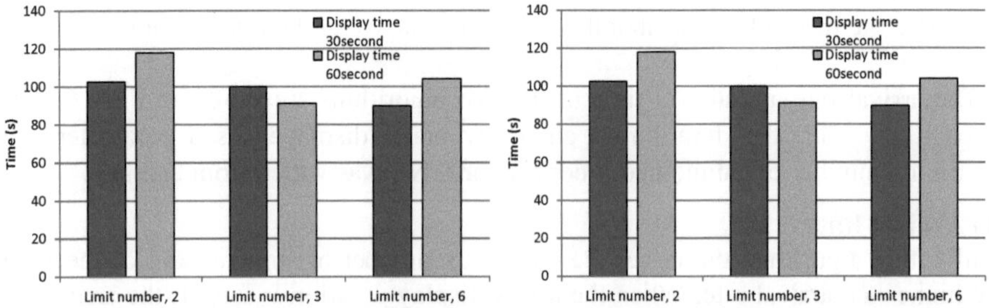

(a) Angle based navigation algorithm (b) Distance based navigation algorithm

Fig. 7. Arrival Time

5 Conclusions

This paper has proposed two navigation algorithms (angle-based navigation and distance-based navigation) for cases where there are multiple signage units and multiple mobile terminals. It has been assumed that multiple mobile terminals communicate with a signage unit simultaneously. We have conducted simulation to evaluate the two algorithms for a case where the maximum number of mobile terminals communicating with a signage simultaneously is limited. It has been found that the angle-based navigation algorithm produces higher arrival rate and shorter arrival time than the distance-based navigation algorithm.

An issue to be addressed in the future is to revise the distance-based algorithm because it currently has an intrinsic problem. It is also necessary to look at security issues. For example, it will be necessary to study countermeasures for cases where a signage unit provides wrong navigation as a result of a malicious user having sent ill-intended data to that signage unit.

References

[1] Waaaaay! A navigate application of distance and direction, `http://app-walker.jp/android-2/76397.html` (accessed on November 14, 2014)
[2] Libelium: Detecting iPhone and Android Smartphones by Wi-Fi and Bluetooth, `http://www.libelium.com/products/meshlium/smartphone-detection` (accessed on May 27, 2014)
[3] iBeacon for Developers - Apple Developer, `https://developer.apple.com/ibeacon` (accessed on August 30, 2014)
[4] Farid, Z., Nordin, R., Ismail, M.: Recent Advances in Wireless Indoor Localization Techniques and System. Journal of Computer Networks and Communications 2013, Article ID 185138, 12 pages (2013), doi:10.1155/2013/185138
[5] Yamaguchi, T., Takami, K.: Estimation of Locations in an Underground Shopping Mall through Inter-Terminal Message Exchange. In: IEEE 9th International Conference on Wireless and Mobile Computing, Networking and Communications (WiMob 2013), October 7-9 (2013)
[6] Takahashi, D., Takami, K.: Costomer derivative method in the underground shopping center using digital signage. Inst. of Electronics Information networks study The 19th Tokyo Branch student meeting for presenting research papers 190, March 1 (2014)
[7] Boswarthick, D., Elloumi, O., Hersent, O.: M2M Basic technical book ETSI Standard theory and system. Rick telecom Co, Ltd. (December 26, 2013)
[8] Tobu hope center, `http://www.tobu-hope.co.jp/` (accessed on December 12, 2012)

MANETs Routing Method for Calls with Predefined Duration in Audio/Video Communications

Hiroyuki Todoroki and Kazumasa Takami*

Graduate School of Engineering, Soka University
1-236 Tangi-cho, Hachiouji-shi, Tokyo, 192-8577 Japan
k_takami@t.soka.ac.jp

Abstract. MANETs (Mobile Ad hoc Networks) are attracting attention as a useful means of communication in times when the communications network infrastructure has been disrupted by a widespread disaster. So far, audio/video services provided on MANETs have not been evaluated adequately. Audio/video communication generates large packet traffic, which rapidly reduces the battery levels of relaying terminals. This results in ongoing communication being interrupted due to a low battery level. This will lead to frequent disconnections and attempts to reestablish network routes at short intervals, resulting in degradation in communication quality. This paper studies a routing method that takes account of remaining battery levels in multimedia communication. Specifically, it proposes a routing method that establishes communication routes for calls with predefined communication duration, using only those terminals that have adequate remaining batter levels. The effectiveness of the proposed routing method has also been confirmed through communication quality evaluation by using a developed SIP_MANET emulator.

1 Introduction

Recently, there has been a growing interest in MANETs (Mobile Ad hoc Networks) [1]-[3] as a means of communication between wireless terminal users in cases where the existing communications network infrastructure has been disrupted by a widespread disaster or where it is difficult to install a communications network infrastructure in the first place. A MANET is a network in which multiple wireless terminals communicate with each other directly, or via other wireless terminals if the two terminals concerned are so far from each other that their radio waves cannot reach each other.

In the Internet, connection-oriented applications (for audio/video multimedia communication) using SIP (Session Initiation Protocol) [4] are gaining in popularity, and many applications have been or are being developed [5]-[7]. Carriers that provide fixed-line IP phone services have adopted SIP [7]. WebRTC Session Controller that can be connected with existing SIP network has also been developed [8].

If mobile terminals have the capability to select a network and switch to it autonomously, it can switch to MANET mode in the event that the conventional communications network infrastructure is devastated by a widespread disaster. This will

* Corresponding author.

K.J. Kim and N. Wattanapongsakorn (eds.), *Mobile and Wireless Technology 2015*,
Lecture Notes in Electrical Engineering 310, DOI: 10.1007/978-3-662-47669-7_6

enable the user to continue audio/video communication that he/she is already familiar with.

However, when audio/video communication uses MANETs, traffic increases, which rapidly reduces the battery levels of relaying terminals. This results in ongoing communication being interrupted due to a low battery level. This will lead to frequent disconnections and then connections of network routes at short intervals, resulting in degradation in communication quality. Existing routing methods that focus on battery levels are MBCR (Minimum Battery Cost Routing) [3] and MMBCR (Min-Max Battery Cost Routing) [3]. These focus only on the battery levels in a route. As a result, they may sometimes leave nodes with a low battery levels within the route. Communication that involves transmission of large traffic, such as audio/video communication, consumes far more battery power that text data communication. Therefore, frequencies of route disconnections due to a low battery level increase.

This paper proposes a routing method that reduces the frequencies of route disconnection due to relaying terminals not having adequate battery levels in MANETs. This is achieved as follows. An application in each terminal calculates the required battery level from the communication duration requested by the user, and chooses a route that is made up of terminals that satisfy the required battery level. Section 2 proposes a routing method that takes account of battery levels in audio/video communication. Section 3 describes the SIP_MANET emulator system that has been developed to evaluate communication quality. Section 4 examines the effectiveness of the proposed architecture through a subjective evaluation of audio/video communication quality. Finally, Section 5 presents the conclusions and remaining issues.

2 Routing Method for Calls with Predefined Communication Duration

Here we propose a new routing protocol based on AODV (Ad hoc On-Demand Distance Vector), which is a reactive protocol. The proposed protocol has been conceived to avoid disconnections of ongoing communication due to a low battery level in audio/video communication, which generates large packet traffic, and to take account of node mobility. The SIP application in each terminal calculates, in advance, the battery level required in the terminal to support the requested duration of communication. It puts information about the required battery level in an RREQ (Route Request) packet, broadcasts the packet, and establishes a route that is made up of terminals that have the required battery level. This can avoid ongoing communication from being disconnected due to a low battery level.

The proposed method determines the required battery level and establishes a route that does not include terminals that do not have the required battery level. If a route is simply established with terminals that have high battery levels, some relaying terminals in the route may not have adequate battery levels, which can result in disconnections of ongoing communication. To avoid this situation, sender, S. puts information about the required battery level in an RREQ, and broadcast it so that terminals with battery levels lower than the required level will not be included in the route to be established. Figure 1(a) shows an example of flooding RREQs that include information about the required battery level. The RREQ sent by Terminal S is flooded by intermediate relaying terminals and reaches receiver, D. In normal AODV, any terminal that has received an RREQ broadcasts it unless the received RREQ is a duplicate of a previously received RREQ. In our proposed method, a terminal that has received an RREQ compares its own battery level with the required battery level shown in the RREQ. If its own battery level is lower than the required battery level, the terminal does not broadcast the RREQ. This ensures that the route to be established does not include terminals whose battery levels are lower than the required level. Figure 1(b) shows an example in which a route is established with due consideration for battery levels. In AODV, as shown by dotted lines, a route with the smallest number of hops is selected without consideration for battery levels. Suppose that a battery level of 60 is required for communication. The route selected in AODV includes a relaying terminal whose battery level is 44, which is lower than the required level. Communication in progress on this route may be disconnected. The proposed routing algorithm establishes a route that does not involve terminals whose battery levels are lower than the required battery level.

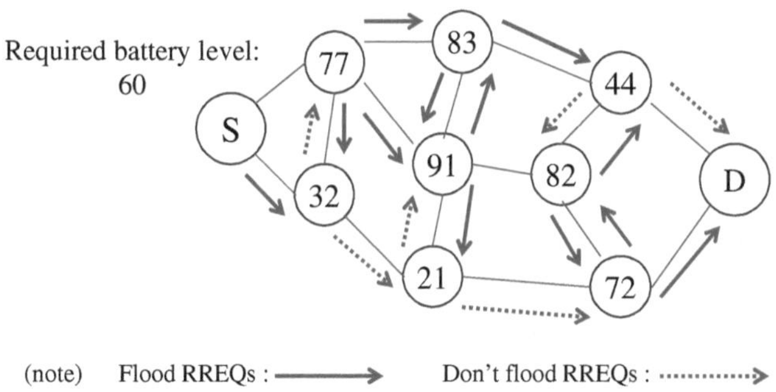

(a) RREQ flooding that takes account of battery levels.

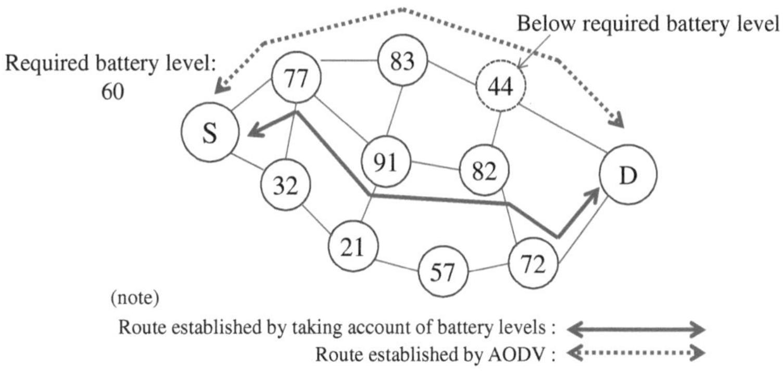

(b) Example of routing that takes account of battery levels.

(note) *S*: sender. *D*: receiver.
The number in each circle indicates the battery level of the terminal concerned.

Fig. 1. Routing method that takes account of battery levels for calls with predefined communication duration

The algorithm for establishing such a route is described below. As shown in Fig.2, let B be a terminal battery level, let N_w, N_k, N_m, and N_p be relaying terminal numbers, and let B_w, B_k, B_m, B_p be the battery levels of these terminals, respectively. Figure 3 shows the RREQ message format that has been extended to implement the proposed routing protocol. The shaded parts in the format are newly added parts.

Step 1: The user of the sender determines the communication duration he/she wants.

Step 2: The battery level required to support that duration of communication is calculated.

Step 3: Sender, S, inserts B_{Nece} into *the required battery level*, and zero into *the minimum battery level in the route* (B_{Mini}) in its RREQ, and broadcasts the RREQ.

Step 4: A node that has received one of the flooded RREQs refers to B_{Nece} and B_{Mini}, and makes the following conditional branching:

Case 1: $B \geq B_{Nece}$ and $B \geq B_{Mini}$

N_w that has received the RREQ compares B_w with B_{Nece} in the RREQ. Since B_w is equal to or greater than B_{Nece}, N_w qualifies as a relaying terminal. It also compares B_w with B_{Mini}. Since B_w is equal to or greater than B_{Mini}, it does not change the value of B_{Mini}. It increments Hop Count by one, and broadcasts the RREQ.

Case 2: $B \geq B_{Nece}$ and $B < B_{Mini}$

Since B_m is equal to or greater than B_{Nece}, N_m that has received the RREQ qualifies as a relaying terminal. Since B_m is smaller than B_{Mini}, it inserts B_m into B_{Mini}. It increments Hop Count by one, and broadcasts the RREQ.

Case 3: $B < B_{Nece}$

Since B_k is smaller than B_{Nece}, N_k that has received the RREQ does not qualify as a relaying terminal. Therefore, it does not broadcast the RREQ.

Step 5: N_p that has received RREQs from multiple terminals refers to RREQ ID in the last received RREQ. If the RREQ ID is the same as an RREQ ID received earlier, N_p does not broadcast the RREQ.

Step 6: The receiver, which ultimately receives the RREQ, refers to B_{Mini} and Hop Count in the RREQ. What it will do thereupon depends on the following conditional branching:

Case 1: In the case of the routing method that gives priority to battery levels;

The receiver selects the route in which B_{Mini} in the received RREQ is the highest. If multiple routes have the same B_{Mini} value, it selects the route with a smaller Hop Count. Go to Step 7.

Case 2: In the case the routing method that gives priority to the number of hops;

The receiver selects the route in which Hop Count in the received RREQ is the smallest. If multiple routes have the same Hop Count value, it selects the route with the highest B_{Mini}. Go to Step 7.

Step 7: When the route is disconnected because a relaying terminal involved in the route has moved, a new route is established based on the applicable routing method.

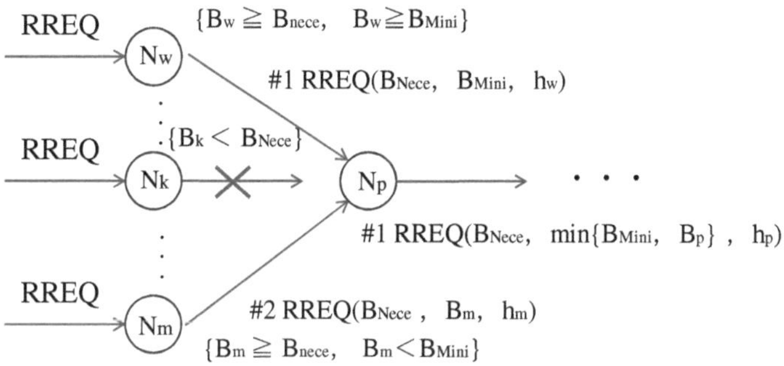

(*Note*) B_{Nece} : Required battery level. B_{Mini} : Minimum battery level in the route. h : Hop count.

Fig. 2. RREQ transmission taking account of battery levels

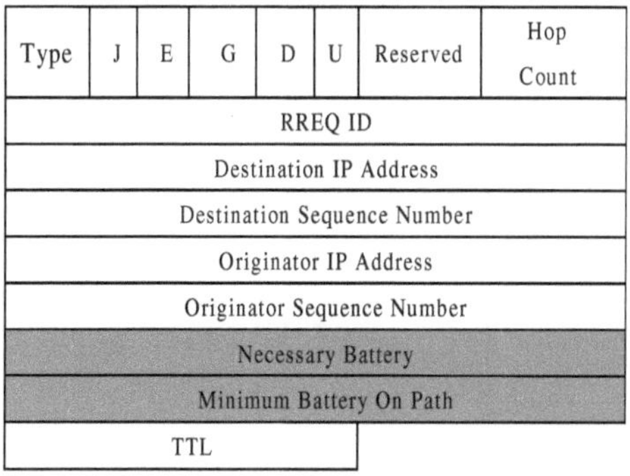

(Note) *Necessary Battery*: the required battery level for relaying terminals. This is calculated based on the requested duration of communication in advance. *Minimum Battery On Route*: the minimum battery level in the route at the time the route is established.

Fig. 3. Extended RREQ packet format

3 SIP_MANET Emulator

We have developed the SIP_MANET emulator [9]. There are open software programs that provide SIP services, including Linphone [10], X-Lite [11] and SIP Communicator [12]. In the current study, we use Linphone, which can communicate without using a SIP server as long as the relevant IP address is known.

Windows displayed by the developed system are shown in Fig. 4. Figure 4 shows console windows of different terminals connected to the SIP_MANET emulator. Packets are sent from the sender to the receiver. We set the required battery level to 60, and set the initial battery level of each terminal to a random value between 20 and

100. In this example, the route from the sender to the receiver goes through either Terminal 4 or Terminal 18. However, since the initial battery level of Terminal 4 is 24, this terminal is excluded from the list of candidate routes. When a terminal receives an RREQ, it compares its own battery level with the required battery level in the RREQ. If its own battery level is lower than the required battery level, the terminal determines that it cannot become a part of any candidate route, and does not flood the RREQ. In the current study, once a terminal has been excluded from the list of candidate routes, its battery level is set to zero in order to facilitate management of terminals. In this example, the route including Terminal 4 is excluded, and the route including Terminal 18, which satisfies the required battery level, is selected for the communication. It was confirmed that the receiver, Terminal 5, received packets correctly.

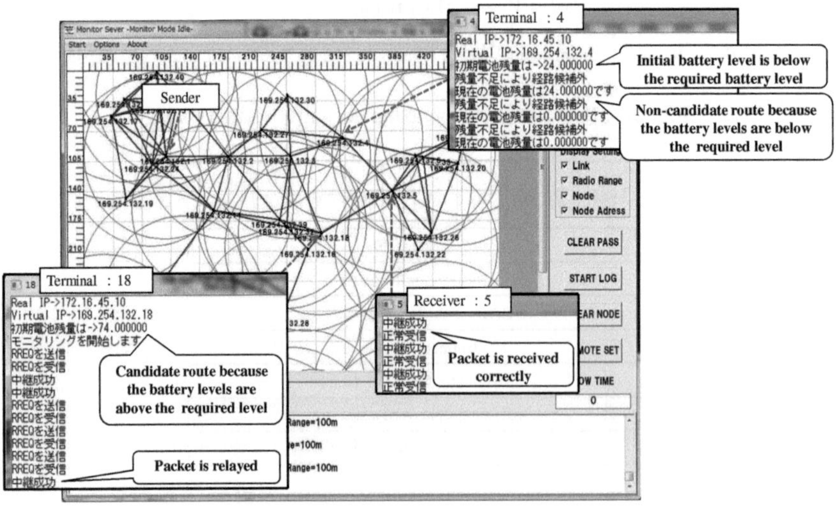

Fig. 4. Communication routes shown on the display of the emulator and a console window on each terminal

4 Evaluation

Using the SIP_MANET emulator, we evaluated audio/video communication under the conditions shown in Table 1. Linphones were used for audio/video communication.

Figure 5 compares the three routing methods in terms of the total length of time in which audio/video communication was carried out correctly. The figure shows that the length of time in which communication was possible was longer when the proposed routing methods were used than when AODV was used. A reason for this is that, as in the case of data communication, the chances of established routes for ongoing communication being disconnected due to depletion of the battery of a relaying terminal in the route were smaller with the proposed routing methods than with AODV.

Figure 6 shows the lengths of time when communication was possible or not possible with the three routing methods for a case where there were 45 nodes. The non-red parts represent the time when communication was not possible. This graph confirms that the proposed routing methods provide long communication time while AODV cannot provide stable communication time because communication interruptions occur frequently.

Table 1. Measurement Conditions

Parameter	Value
Routing	Proposed methods and AODV
Number of nodes	20 to 45
Node mobility area	500×500 (m^2)
Area covered by radio wave	Radius: 100 m
Node moving speed	6.4 (km/h)
Node mobility model	Random way point
Required battery level	60
Initial battery level	20 to 100 (random)
Power consumption timing	Time of reception or transmission
Consumed power per packet	During media communication: 0.001
Requested duration of communication	3 minutes
Number of experiments	10

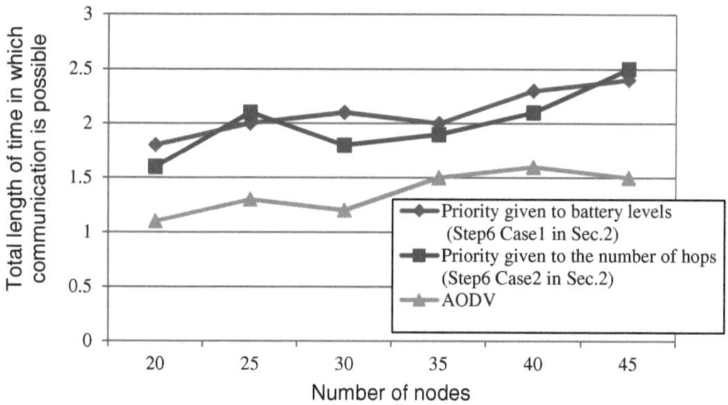

Fig. 5. Total length of time in which audio/video communication was possible in AODV and the proposed routing methods

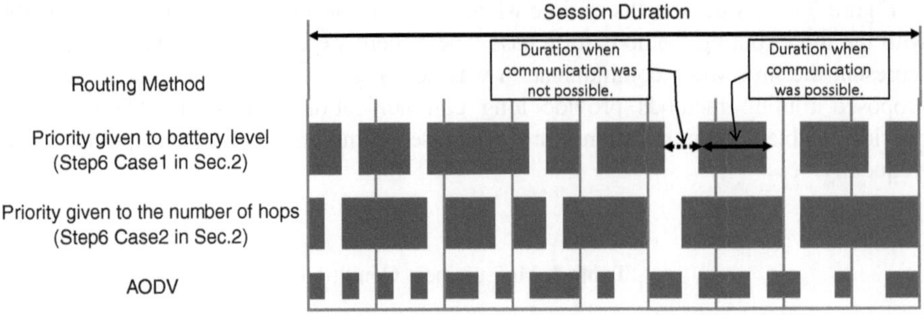

(Note) *Non-red part*: duration when audio/video communication was not possible although the session is still held

Fig. 6. Length of time when communication was possible or not possible (for a case where there are 45 nodes)

5 Conclusions and Future Issues

A MANET is a useful means of communication in times when the communications network infrastructure has been disrupted by a widespread disaster. The paper has proposed a routing method that takes account of battery levels of terminals so that multimedia communication can be used effectively. It has also evaluated the length of time in which communication is possible and the probability of successfully establishing a route. We have confirmed that the proposed routing method can provide greater availability of communication than existing AODV. It has been found that a practical level of grade of service can be provided if the assumed duration of communication is up to 3 minutes, which can be the case with communication used to inquire about the safety of people in a disaster area.

It will be necessary to carry out evaluation for communication that involves multiple sessions simultaneously in order to confirm the effectiveness of the proposed routing method in a real environment.

References

1. Kumar Sarkar, S., Basavaraju, T.G., Puttamadappa, C.: Ad Hoc Mobile Wireless Networks, Principles, Protocols and Applications. Auerbach Publications, Boston (2007)
2. Varaprasad, G., Wahidabanu, R.S.D., Venkataram, P.: An efficient resource allocation scheme for multimedia applications in MANET. Journal of Network and Computer Applications 31(4), 577–584 (2008)
3. Toh, C.K.: Maximum Battery Life Routing to Support Ubiquitous Mobile Computing in wireless Ad Hoc Network. IEEE Communication Magazine, 138–147 (June 2001)
4. Wauters: Skype Opens Up To SIP. Finally Eyes Enterprise Customers, The Way It Should, `http://techcrunch.com/2009/03/23/skype-opens-up-to-sip-finally-eyes-enterprise-customers-the-way-it-should/` (accessed on April 25, 2012)

5. Banerjee, N., Acharya, A., Das, S.K.: Enabling SIP-based sessions in ad hoc networks. Wireless Networks 13(4), 461–479 (2007)
6. Leggio, S., Manner, J., Hulkkonen, A., Raatikainen, K.: Session initiation protocol deployment in ad-hoc networks: a decentralized approach. In: Proceedings of the International Workshop on Wireless Ad-Hoc Networks (IWWAN 2005), London, UK (May 2005)
7. SIP social.net, http://www.sipsocial.net/sipsocial-apps/sip-4g-voip (accessed on April 25, 2012)
8. Oracle communications webrtc session controller, http://www.oracle.com/us/products/applications/communications/web-rtc-session-controller/overview/index.html (accessed on October 24, 2014)
9. Todoroki, H., Kagoshima, T., Kasamatsu, D., Takami, K.: Implementation of a Peer-to-Peer-type SIP Client Application on a MANET Emulator. In: TENCON (IEEE Region 10 Conference), November 19-22 (2012)
10. Linphone – open source video SIP phone for desktop & mobile, http://www.linphone.org/ (accessed on May 31, 2012)
11. X-Lite, http://www.counterpath.com/x-lite.html (accessed on May 31, 2012)
12. SIP Communicator (Jitsi), https://jitsi.org/ (accessed on May 31, 2012)

Enhanced Adaptive Modulation and Coding (AMC) Technique Using ILDPCA Coders for MIMO Systems

J. Sofia Priya Dharshini[1,*], M.V. Subramanyam[1], and K. Soundararajan[2]

[1] ECE, RGMCET, Nandyal, Kurnool Dist., Andhra Pradesh, India, 518501
[2] ECE, KITES, RangaReddy Dist., Telagana, India, 509217
sofi_prida@yahoo.co.in,
{mvsraj,soundararajan_jntucea}@yahoo.com

Abstract. AMC has been adopted at the physical layer of several standard e, g 3GPP, 3GPP2, HIPERLAN/2, IEEE802.11a, IEEE 802.15.3 and IEEE 802.16. MIMO technology in coordination with Adaptive Modulation and Coding (AMC) scheme further enhances spectral efficiency by adjusting transmission parameters to the channel condition while satisfying a target error performance. Consequently, the combined MIMO-AMC technology is a promising solution to offer high throughput for next generation wireless systems. The system still has limitations in terms of channel estimation error, periodic performance drop, channel correlation resulting in reduced throughput and packet loss rate A cross layer design using Improved Low Density Parity Check Code (ILDPCA) in AMC on the physical layer and Incremental redundancy Hybrid ARQ (IR-HARQ) in the data link layer is proposed. ILDPCA coder in AMC enhances the performance of the system in terms of bit error rate,packet loss rate and spectral efficiency.

Keywords: Improved Low Density Parity Check Code (ILDPCA), Incremental redundancy Hybrid ARQ (IR-HARQ), Adaptive Modulation and Coding (AMC), Multiple Input Multiple Output (MIMO).

1 Introduction

MIMO systems found the way into several standards for future wireless communication systems, especially in Wireless Local Area Networks (WLAN) and cellular networks such as IEEE 802.11, 802.16 and the 3rdGroup Partnership Project (3GPP) [1]. When Adaptive Modulation and Coding (AMC) scheme is used along with MIMO technology, it increases spectral efficiency by amending transmission parameters to the channel condition but still satisfying the target error performance [2]. The link adaptation techniques like adaptive modulation coding could dramatically improve the spectral utilization efficiency and throughput of wireless networks [3][4]. Link-adaptation means adaptation of the code rate, the modulation constellation, and optionally, the power-allocation. The latter two are often named as bit- and power loading. The objective of link-adaptation is to estimate the channel at the receiver and to feed this estimation back

* Corresponding author.

K.J. Kim and N. Wattanapongsakorn (eds.), *Mobile and Wireless Technology 2015*,
Lecture Notes in Electrical Engineering 310, DOI: 10.1007/978-3-662-47669-7_7

to the transmitter so to adapt the transmission scheme to the channel conditions. Various coding schemes,Binary Phase Shift Keying (BPSK), Quadrature PSK (QPSK), 8-PSK, 8-QAM (Quadrature Amplitude Modulation), 16-PSK, 16-QAM, 64-PSK, 64-QAM etc. are used [5].However the exact number of environmental parameters like signal energy, noise variance, channel state information and other subtle factors like quantization error, made it hard to tune the transmission parameters optimally (even near optimal) [4]. Erik Johannesson et al [6] proposed the joint design of an optimal linear MIMO encoder-decoder pair of parallel Gaussian channels that can be formulated as a convex optimization problem followed by a matrix spectral factorization. It concentrated on a mixed norm minimization problem, in which the relative weight of the two norms is determined by the maximum transmission power. However, it was not sure that this technique would be applied to all other structures like feedback loops. Mahendra Babu D.S et al [7] presented a memory-efficient and faster convolution encoding and adaptive Viterbi decoding implementation technique for MIMO-OFDM communication systems on an FPGA. The MIMO-OFDM implementation required to deploy the minimum required on-chip memory. The I/O rate for MIMO-OFDM based communication systems are doubled for 2×2 MIMO systems in the proposal.

Gaurav Thakur [8] combined the processes of decoding and demodulation with simple, DSSS-based and repetition like coding schemes which could confer significant advantages in high noise environments. These results are generalized to broader classes of short-length block codes or other modulation schemes, as well as other demodulation approaches such as those that output soft symbol decisions and/or have observation intervals of multiple data bits. Chao Yu and Gaurav Sharma [9] presented improved constructions for Low-Density Parity-Check Accumulate (LDPCA) codes, which are rate adaptive codes commonly used for distributed source coding (DSC) applications. The proposed constructions mirror the traditional LDPCA approach; higher rate codes are obtained by splitting the check nodes in the decoding graph of lower rate codes, beginning with a lowest rate mother code. However, the proposed constructions introduce non-uniform splitting of the check nodes at higher rates. Sofia et al [10] proposed cross layer design with AMC in the physical layer and IR-HARQ in the data link layer.Convolution codes are used for channel coding. The reliability of the system can be enhanced with high performance channel coding methods.Improved low density parity check coders is used as a channel coding technique instead of convolution coder and Viterbi decoder.In the second section, the proposed system using ILDPCA is presented. Section 3 presents the ILDPCA encoder and decoder.In section 4 and section 5, result analysis and conclusion is presented.

2 Proposed System

The proposed cross layer approach connects physical layer and data link layer to enhance the performance of MIMO network.AMC controller is used in the physical layer in conjunction with IR-HARQ in the data link layer of the system. AMC controller adaptively changes the modulation and code rate based upon the prevailing channel conditions. The channel is estimated at the receiver and is transmitted to the transmitter section using feedback channel.IR-HARQ is utilized at the data link layer to regulate packet retransmissions. At each retransmission, different code words z_i are forwarded in IR-HARQ.MIMO system with Ntx number of transmitting antennas and Nrx number of receiving antennas as shown in Fig.1. Improved Low-Density Parity Check Accumulate Codes (ILDPCA) encoder is used at the transmitter. By means of MIMO fading channels, the coded symbols are forwarded at the physical layer on a frame by frame fashion.

The proposed ILDPCA encoder has three stages. The first stage is an invertible linear transformation, s = Vx of the source data sequence x, where V is a L×L non-singular sparse binary matrix. The second stage is a rate 1 accumulator that takes the length L sequence ,**s** $\stackrel{def}{=}[s_1, s_2, \ldots, s_L]^T$ and generates the length L sequence $c = [c_1, c_2, \ldots c_L]^T$, where $c_i = \sum_{j=1}^{i} s_j$ The third stage permutes the sequence c, using a permutation π of the indices [1,2,...,L], to obtain a sequence $\mathbf{p} = [p_1, p_2, \ldots, p_L]^T$. **s** and **c** are referred to as syndrome and accumulated syndrome sequences, respectively. A discrete set of *N* possible rates is enabled by the rate scalability, where the integer parameter *N* is a factor of the block length *L* so that $M = L/N$ is an integer. The encoder first sends *M* syndromes to the decoder and responds with *M* additional syndromes in response to each decoder request for additional bits. The decoder's attempt to recover **x** based on already received data fails. The process continues until decoding succeeds, which is usually verified using a checksum of the source message that is independently communicated to the decoder, resulting in a (negligible) overhead. The rate-adaptively offers the discrete set of *N* rates ($1/N, 2/N, \cdot \cdot \cdot, (N-1)/N, 1$). The permutation π that maps **c** to **p** [9]. This generated signal is drawn through the encoder and the information is modulated and transmitted using STBC transmitter. The input signal is generated randomly as.

$$X_{ij}=[x(1,1),x(1,2),\ldots] \tag{1}$$

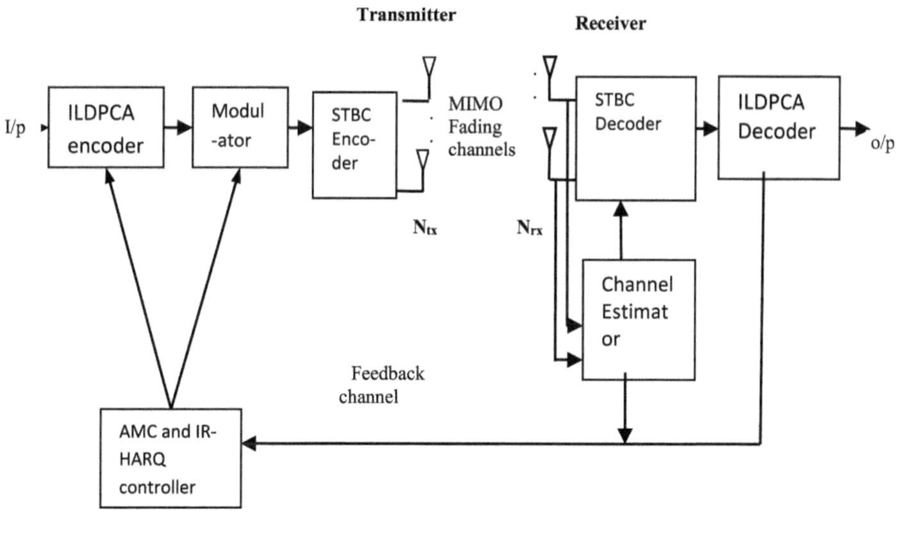

Fig. 1. Proposed System Architecture

The diversity order is described as:

$$D \overset{\Delta}{=} N\,tx \;\; N\,rx \tag{2}$$

The receiver computes the SNR and sends back to the AMC controller. The controller selects a suitable Modulation and Coding Scheme (MCS) for the next transmission. If ds is the data sequence, it is encoded into z_i code words where $i = 0,1,\ldots,z\text{-}1$.Therefore more than one code word z_i may contain the part of information in ds. The technique makes use of ILDPCA for encoding *the* mother code, which is punctured to generate the z_i. Diverse z_i may enclose common systematic or parity bits of the mother code. Here, Block flat fading is considered for the channel remains constant during the transmission of one symbol. If the channel is known only at the receiver, the receiver tries to decode the received data. If no error detected or the decoder corrects the errors, then the successful reception is indicated to the transmitter. Then, the system progresses on transmitting a new data sequence ds. Upon a failure, a retransmission is invoked. Previously transmitted data is tracked at the receiver in order to associate the decoding, since the channel is known only at the receiver.

2.1 IR-HARQ

Incremental Redundancy Hybrid Automatic Repeat Request (IR-HARQ) is used in the data link layer to reduce the number of retransmissions and meet the delay constraints. A Maximal Ratio symbol Combining (MRC) approach is used in a Maximum-Likelihood (ML) receiver. When Maximum Ratio symbol Combining (MRC) is used along with a block-fading channel, the received symbols are collaborated as [12],

$$\tilde{Y}_0 = \sum_{i=0}^{n-1} h_i^* Y_i = \tilde{h}_0 x_0 + \sum_{i=0}^{n-1} h_i^* n_i = \tilde{h}_0 x_0 + \tilde{z}_0 \tag{3}$$

Where, $\tilde{h}_0 = \sum_{i=0}^{n-1} h_i^* h_i$ and n stands for the number of transmissions. To measure the Log-Likelihood Ratios (LLRs), the ML receiver uses $Y_0^{'} = \frac{1}{\sqrt{\tilde{h}}}\tilde{Y}_0$ and $\sqrt{\tilde{h}}$, which will be transmitted to the decoder so as to estimate the information sequence *ds*. At each retransmission, different code words z_i are forwarded in IR-HARQ.

3 ILDPCA

Using MIMO fading channels, the coded symbols are forwarded at the physical layer on a frame by frame fashion subsequently using Improved Low-Density Parity Check Accumulate Codes (ILDPCA) technique. This code is defined by the matrix V along with the permutation π .

Let $V^{(1)}$ be the lowest rate mother code.
Let q be the number of requests received by the encoder.
where q = 1 for initial transmission.
Let q_{up} be the upper bound rate value; where $2 \leq q_{up} < N$ is the integer value.
Let τ be the transmission range.

The steps involved in the construction of ILDPCA code from the lowest rate mother code using splitting technique are as follows:

1. Initially code is constructed by selecting the parity check matrix $V^{(1)}$ for the mother code.

   Each row in $V^{(1)}$ is the sum of a selection of rows from V. The total number of edges in the decoding graph G(V) is preserved in the decoding graphs $\{G(V^q)\}_{q=1}^{N-1}$

   Note: There is no overlap between the non-zero entries in the selected rows.

2. As per scheduling order defined by the permutation π, the matrix rows are split to generate $V^{(q)}$ from $V^{(q-1)}$ for $2 \leq q \leq N$.

3. When splitting the selected row α^T in $V(q)$ into two rows σ^T and τ^T in $V^{(q+1)}$, non-zero entries are partitioned equally and randomly between σ^T and τ^T.

$$\alpha^T = V^{(q-1)}_{(q-1)(i-1)+d}$$

$$\sigma^T = V^{(q)}_{q(i-1)+d}$$

$$\tau^T = V^{(q)}_{q(i-1)+d+1}$$

   Splitting of rows involves the following scenario:

   - If X rows $V^{(q)}$ are to be split by the transmission order π, then β X are randomly selected. where β is pre-defined code parameter. Each selected row α^T is split into two rows σ^T and τ^T in $V^{(q+1)}$

   Here, σ^T contains the degree as 2 and τ^T contains the degree as $(\varepsilon_V(\psi)-2)$ where $\varepsilon_v(b)$ = hamming weight of the binary vector b.

- Similarly ,when the remaining (1- β)X are designated for splitting, each selected row α^T is split into two rows σ^T and τ^T with degrees 3 and $(\varepsilon_V(\psi)-3)$ respectively.

For the splits into degree 2 (3) nodes, 2 (3) of the non-zero entries of σ are randomly selected for allocation to τ^T .

1. $V^{q}_{q(i-1)+j} \leftarrow V^{(q-1)}_{(q-1)(i-1)+j}$ for $1 \leq j < n$
2. $V^{q}_{q(i-)+j} \leftarrow V^{(q-1)}_{(q-1)(i-1)+j-1}$ for $(n+1) < j \leq q$
3. The above steps are repeated until the code is generated with lowest rate mother code $V^{(1)}$.

These Rate adaptive codes are commonly used for distributed source coding (DSC) applications and for several DSC applications such as distributed video coding and image authentication.

4 Simulation Results

The receiver computes the signal to noise ratio (SNR). The complex modulated symbols (R) are mapped by STBC encoder into T_N orthogonal complex symbol sequences of length *nS*. These mapped symbol sequences are forwarded concurrently by T_N. As a result, the coding rate (C_R) of a STBC can be given as,

$$C_R = \frac{R}{nS} \tag{4}$$

At the receiver, SNR can be calculated as follows,

$$\gamma = \frac{AvgP}{\sigma^2}\|H\|_F^2 = \frac{tP}{\sigma^2 T_N C_R}\|H\|_F^2 = \frac{\overline{\gamma}}{T_N C_R}\|H\|_F^2 \tag{5}$$

Here, *tP* is the total transmission power transmitted at the antennas of T_N for every symbol duration. The average pseudo SNR ($\overline{\gamma}$) is given as,

$$\overline{\gamma} = \frac{tP}{\sigma^2} \tag{6}$$

The simulation parameters used are as shown in Table 1.

Table 1. Simulation Parameters

No	Parameters	Values
1	Number of transmitters (Nt)	2
2	Number of receivers (Nr)	2
3	Modulation	2,4,16,64
4	Number of packets	100
5	Coding technique	ILDPCA
6	SNR(dB)	0:2:20
7	Channel	Rayleigh Fading
8	Gain vector	[0 -3 -6 -9]

The proposed system is simulated using the parameters specified in Table 1.Performance plots of the proposed system are presented in terms of BER,PER,Average rate and spectral efficiency Fig. 2 shows the bit error rate performance of the system using ILDPCA and without using ILDPCA. It is evident that the bit error rate is reduced by a considerable factor.Fig.3 shows the average rate performance of the proposed design.Around 70 % of improvement in the data rate is observed for the channel with SNR of 4 dB.

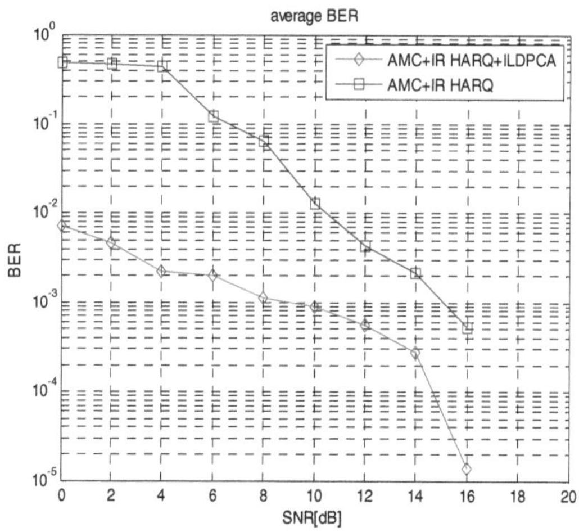

Fig. 2. Average BER performance of the proposed system

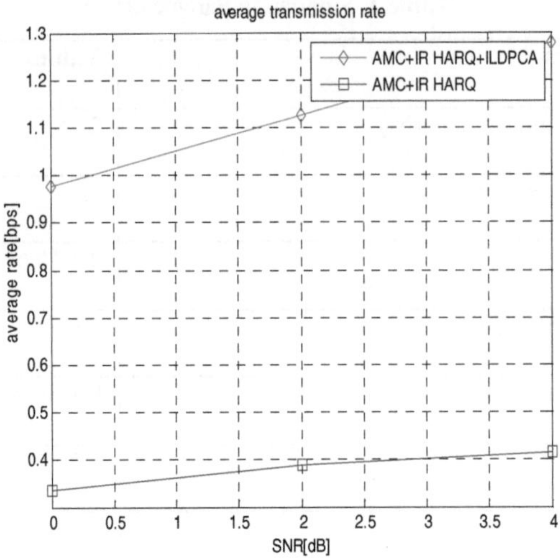

Fig. 3. Transmission Rate performance using ILDPCA

Fig.4.shows the spectral efficiency plot of the system using ILDPCA.Significant improvement in the spectral efficiency is observed in the system with ILDPCA based AMC. In poor channel conditions also,it is evident that the proposed system has an acceptable spectral efficiency.The system has better performance in terms of reliability of transmission.Fig.5 shows that the Packet loss rate is reduced by 80% at 6 dB for the system using ILDPCA.

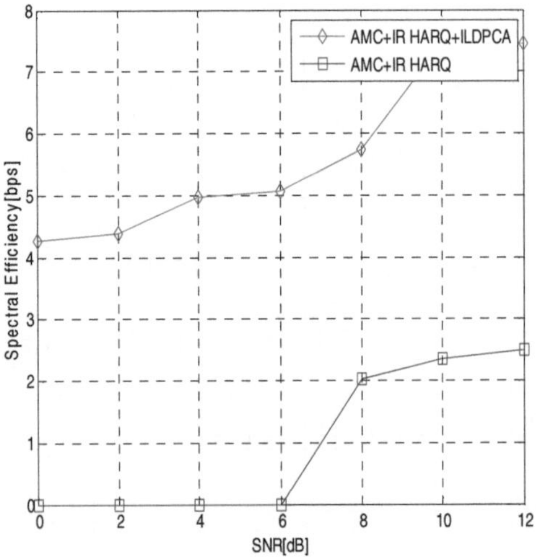

Fig. 4. Average Spectral efficiency using ILDPCA

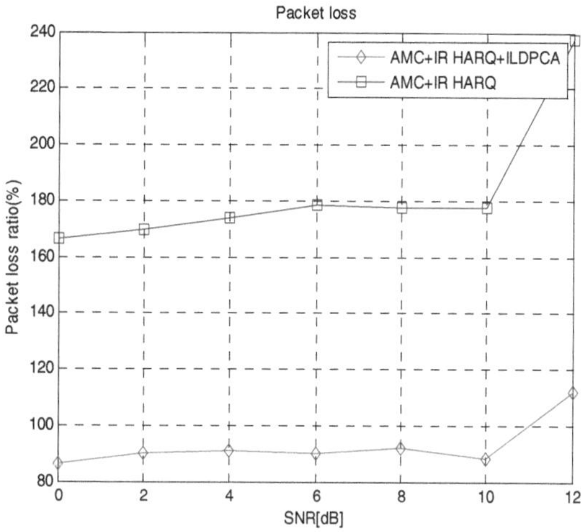

Fig. 5. Packet loss rate performance for the system with ILDPCA

5 Conclusion

Cross layer design using ILDPCA based AMC in the physical layer and IR-HARQ in the data link layer is proposed. ILDPCA codes are used for several DSC applications such as distributed video coding and image authentication.ILDPCA in the proposed cross layer design enhances the performance of the system in terms of reliability,throughput and spectral efficiency.Simulations results are presented in the paper.In real time systems ,the proposed ILDPCA encoder and decoder can be designed for a Xilinx Virtex II XC2V3000 FPGAusing VHDL and System Generator, also from Xilinx.Hence it is evident that AMC with ILDPCA is more suitable for the present wireless communication systems that requires a better QOS.

References

1. Lien, S.-Y., Chen, K.-C., Lin, Y.: Toward ubiquitous massive accesses in 3GPP machine-to-machine communications. IEEE Communications Magazine 49(4), 66–74 (2011)
2. Zhou, S., Zhang, K., Niu, Z., Yang, Y.: Queuing Analysis on MIMO Systems with Adaptive Modulation and Coding. In: IEEE International Conference on Communications (ICC 2008) (2008)
3. Modi, B., Olabiyi, O., Annamalai, A., Vaman, D.: Improving the Spectral Efficiency of Adaptive Modulation in Amplify-and-Forward Cooperative Relay Networks with an Adaptive ARQ Protocol. In: The IEEE Globecom Proceedings (2011)
4. Yun, S., Caramanis, C.: Reinforcement Learning for Link Adaptation in MIMO-OFDM Wireless Systems. In: The IEEE Globecom Proceedings (2010)
5. Saad, W.K., Ismail, M., Nordin, R.: Survey of Adaptive Modulation Scheme in MIMO Transmission. Journal of Communications 7(12) (December 2012)

6. Johannesson, E., Ghulchak, A., Rantzer, A., Bernhardsson, B.: MIMO Encoder and Decoder Design for Signal Estimation. In: Proceedings of the 19th International Symposium on Mathematical Theory of Networks and Systems – MTNS, Budapest, Hungary, July 5-9 (2010)
7. Mahendra Babu, D.S., Vinutha, M.R., Uma, C.: Design and Implementation of MIMO-OFDM using Encoding and Decoding techniques on FPGA. International Journal of Scientific & Engineering Research 5(6) (June 2014)
8. Thakur, G.: Spread Spectrum Codes for Continuous-Phase Modulated Systems. IEEE Transactions on Communications 62(3) (March 2014)
9. Yu, C., Sharma, G.: Improved Low-Density Parity Check Accumulate (LDPCA) Codes. IEEE Transactions on Communications 61(9) (September 2013)
10. Sofia Priya Dharshini, J., et al.: Adaptive Modulation and Coding With Incremental Redundancy Hybrid ARQ in MIMO Systems: A Cross Layered Design. Int. Journal of Engineering Research and Applications 3(5), 503–507 (2013)
11. Thakur, G.: Spread Spectrum Codes for Continuous-Phase Modulated Systems. IEEE Transactions on Communications 62(3) (March 2014)

A Weighted T2F Scheme for WLANs

Huan Zhang, Qinglin Zhao*, Pei Dang, Hongning Dai, and Zhijie Ma

Faculty of Information Technology, Macau University of Science and Technology, Avenida Wei Long, Taipa, Macau, China
huan@buu.edu.cn, {qlzhao,pdang}@must.edu.mo, hndai@ieee.org, mazhijie0000@hotmail.com

Abstract. WiFi network is known to be wasteful because WiFi performs the time-domain contention and often forces channel to remain idle. The time to frequency (T2F) protocol has recently been proposed to perform the frequency-domain contention and it has shown that T2F can reduce the contention overheads effectively. However, T2F just provides fair channel access. In this paper, we proposed a novel weighted T2F (WT2F) design that can provide weighted channel access. We then study the proportional fairness of WT2F. Simulation results validate that the proposed design is very effective and the proposed performance model is very accurate.

Keywords: Weighted, Frequency Contention, WLANs.

1 Introduction

IEEE 802.11 (802.11) wireless LANs (WLANs) have been widely deployed around the world. However, its channel efficiency has received a great attention. The 802.11 networks perform channel contention in time domain. In 802.11, each node must waits for a random time, before packet transmission. When multiple nodes are simultaneously backing off, the channel must remain idle, naturally leading to under-utilization. It has been shown in [5] that more than 30% reduction in throughput due to backing off.

Recently, Ref. [2] proposed a time to frequency (T2F) protocol. In T2F, nodes perform channel contention in the frequency domain, instead of the time domain. T2F arbitrates channel contention in two slots. In slot 1, each node signals on one subcarrier randomly chosen from a pool of subcarriers, and listens to this subcarrier pool at the same time by employing a second antenna. Then each node can determine the winner, who signals on the smallest subcarrier. In slot 2, the nodes choosing the smallest subcarrier perform the 2nd-round frequency-domain contention. The winner transmits packet in next slot. By limiting the channel contentions to two slots, T2F shortens the contention time greatly, compared to the 802.11.

In T2F, each node uniformly selects a subcarrier from the same subcarrier pool. As a result, each node has the same channel access opportunity. However, in reality, different applications have different QoS requirements. For example, Voice packet

* Corresponding author.

K.J. Kim and N. Wattanapongsakorn (eds.), *Mobile and Wireless Technology 2015*,
Lecture Notes in Electrical Engineering 310, DOI: 10.1007/978-3-662-47669-7_8

should have more stringent delay requirement than data packet and therefore should be assigned a higher transmission opportunity. Clearly, T2F cannot fulfill the QoS requirements of real-time applications.

In this paper, we propose a weighted T2F (WT2F) protocol. In WT2F, different nodes choose subcarriers from different subcarrier pools; since subcarrier pools are in different subcarrier ranges, nodes will have weighted opportunities to win the channel. Our contributions are summarized as follows:

- We design a novel WT2F protocol that implements weighted frequency contention.
- We propose a theoretical model to study the proportional fairness of WT2F.
- We validate the proposed model via extensive simulations.

The rest of this paper is organized as follows. In Section 2, we do overview the T2F protocol. Section 3 designs our WT2F protocol. Section 4 theoretically analyzes the performance of WT2F. Section 5 presents simulation results to validate the proposed model. Section 6 concludes this paper.

2 T2F

In this section, we present the basic ideas and the drawbacks of T2F [2].

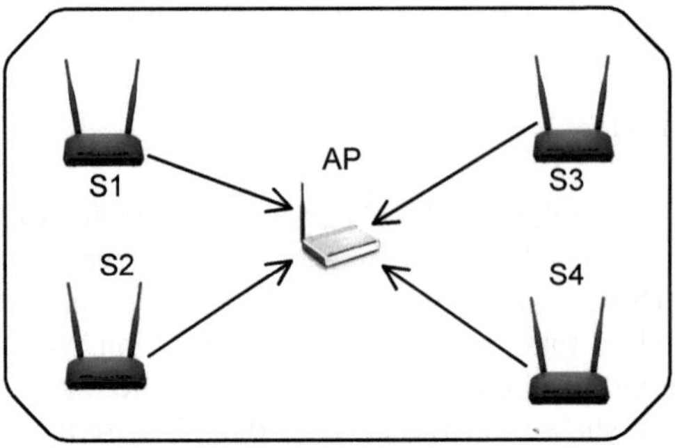

Fig. 1. A star-topology WLAN for T2F

Basic Ideas: T2F is a scheme that provides fair channel access via the contention of the frequency domain for WLANs.

In T2F, each node has two antennas: one for normal data transmission and another for listening to channel. T2F uses the OFDM-based physical layer techniques. In OFDM, the whole channel is divided into L subcarriers (L=52 in 802.11a/g).

Consider a star-topology WLAN illustrated in Fig. 1, where each node Si, i=1,2,3,4 can hear each other and contend for channel for data transmission. With the help of Fig. 2, we now explain the data transmission process in T2F, In T2F, each node first senses channel idle for a DIFS time, then performs 2-round channel contention in two consecutive slots (i.e., R1 and R2 in Fig. 2), and finally execute the data transmission.

In Round 1 contention, each node signals on one subcarrier (via the transmit antenna) randomly chosen from a pool of subcarriers, and listens to this subcarrier pool at the same time by employing the listening antenna. T2F nodes treat the subcarriers as integer numbers. Then each node can determine the winner, who signals on the smallest subcarrier.

In Round 2 contention, the nodes choosing the smallest subcarrier perform the frequency-domain contention in the same way as Round 1 contention.

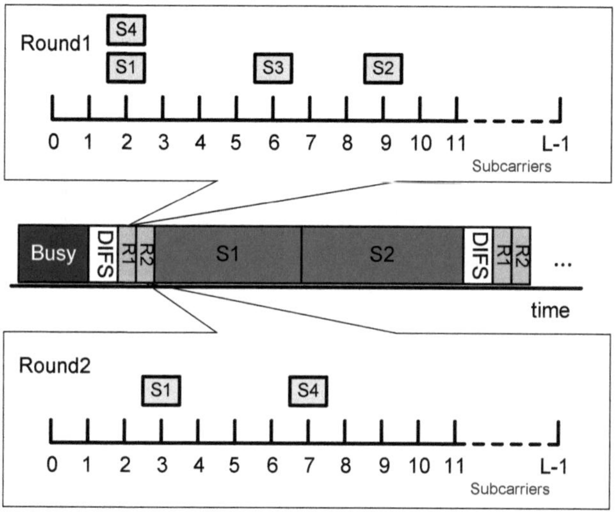

Fig. 2. Two-round contention in T2F

In the example of Fig. 2, in Round 1 contention, S1 and S4 select No.2 subcarrier, S3 and S2 select No.6 and No. 9 subcarriers, respectively. Then S1 and S4 win because their selected subcarrier is minimum. In Round 2 contention, S1 and S4 select No. 3 and No. 7 subcarriers, respectively. After that, S1 and S4 transmit data sequentially. It has been pointed out in [2] that the collision probability (i.e., the probability that multiple nodes select the same subcarrier) in 2-round contention is negligible.

Drawbacks: In T2F, each node uniformly selects a subcarrier from the same subcarrier pool. Therefore, each node has the same channel access opportunity. However, in reality, different applications have different QoS requirements. For example, Voice packet should have more stringent delay requirement than data packet and therefore should be assigned a higher transmission opportunity. Clearly, T2F cannot fulfill the QoS requirements of real-time applications.

3 Weighted T2F

Basic Ideas: In this section, we present the weighted T2F (WT2F). In WT2F, different nodes choose subcarriers from different subcarrier pools; if subcarrier pools are in different subcarrier ranges, nodes will have weighted opportunities to win the channel. When all subcarrier pools are the same, WT2F reduces to T2F.

WT2F Design: In the WT2F design, for simplicity, we only consider two priorities: high priority (HP) and low priority (LP). Each node behaves exactly like that in T2F, except that each node is assigned to HP or LP. For example, in Fig. 3, S1 and S2 are HP nodes, whereas S3 and S4 are LP nodes. HP nodes choose subcarrier numbers in [0, S -1], while LP nodes choose subcarrier numbers in [0, L -1], where S≤L, as illustrated in Fig. 4.

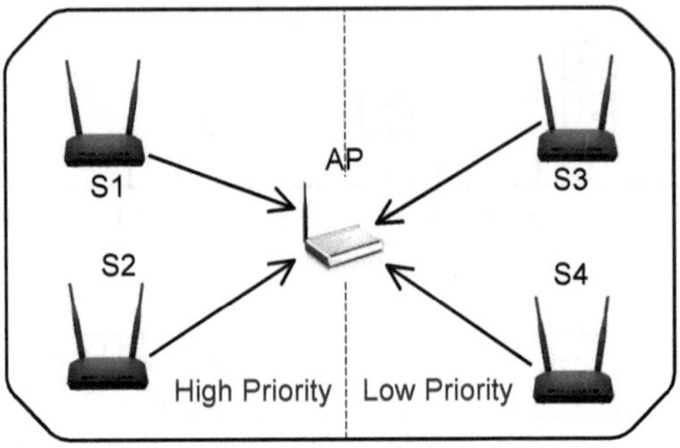

Fig. 3. A star-topology WLAN for WT2F

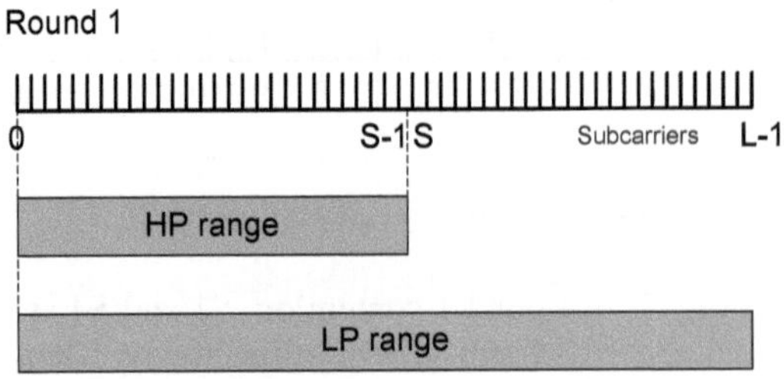

Fig. 4. HP and LP subcarrier ranges for WT2F

The data transmission process in WT2F is exactly like that in T2F, except that in Round 1 contention, HP nodes choose subcarrier numbers in a small range and LP nodes choose subcarrier numbers in a large range. In the example of Fig. 3, HP nodes (i.e., S1 and S2) choose a smaller subcarriers with higher probability than LP nodes (i.e., S3 and S4), so as to achieve weighted channel contention.

4 Performance Analysis

In this section, we study the proportional fairness, the collision problem, and overheads of WT2F.

4.1 Proportional Fairness

In the following, we first present the throughput formula of each node, and then study the proportional fairness by characterizing the relationship among the design parameters (such as S and L) and the number of HP and LP nodes.

Throughput Formula. According to the design of the WT2F, the transmission procedure can be divided into three phases: Firstly, the nodes sense the channel for DIFS interval. Secondly, all the competing nodes start frequency contention within two-Rounds. Finally, the winner nodes transmit the packet. To simplify the analysis, we ignore the collision in the second round since the collision probability is much small [2].

Let T_{DIFS} be the DIFS interval time. Let T_C be the period of two rounds contention. Let T_{DATA} be the transmission period for each packet. In addition, we assume there are n nodes effective transmissions in a contention, $\overline{n}$ is mean value of n. So the system throughput can be expressed as

$$Thp = \frac{P_{DATA} \times \overline{n}}{T_{DATA} \times \overline{n} + T_C + T_{DIFS}} \tag{1}$$

Where P_{DATA} is the packet size for each node. Based on (1), we then express per-node throughput. For the arbitrate node i, assume that the successful transmission probability is P_i, then the throughput for node i is

$$Thp_i = P_i \times \frac{P_{DATA}}{T_{DATA} \times \overline{n} + T_C + T_{DIFS}} \tag{2}$$

From (2), we clearly that the throughput of node i would be different when node i obtains the different successful transmission probability Pi. Next, we try to achieve the proportional fairness by setting different Pi.

Proportional Fairness between Two Nodes. For simplicity, we start our study on proportional fairness from two nodes (only one node in each class). Assume that node A and node B are the HP and LP node, respectively. Node A selects a random subcarrier from the subcarrier pool [0, S-1], while node B selects from [0, L-1], as shown in Fig.4.

Let P_A' and P_B' be the probability that node A and node B choose the arbitrary subcarrier from their subcarrier pool, respectively. Then we have $P_A' = 1/S$, $P_B' = 1/L$. Node A or node B would win the contention if their selected subcarrier number is the minimum. So the successful transmission probability of node A and node B are expressed as (3) and (4), respectively.

$$P_A = P_A' \sum_{i=0}^{S-1} (L - i) P_B' \tag{3}$$

$$P_B = P_B' \sum_{i=0}^{S-1} (S - i) P_A' \tag{4}$$

The ratio of probability of node A and B is below:

$$\frac{P_A}{P_B} = \frac{P_A' \sum_{i=0}^{S-1} (L - i) P_B'}{P_B' \sum_{i=0}^{S-1} (S - i) P_A'} = \frac{2L - S + 1}{S + 1}$$

According to this simple case, next, we extend our analysis from two nodes to multiple node.

Proportional Fairness among Multiple Nodes. Assume that the number of HP nodes is N1, and the number of LP nodes is N2. All the HP nodes select the subcarrier from the same subcarrier pool [0,S-1], and all the LP nodes select the subcarrier from [0,L-1]. Then the ratio of any HP node A to any LP node B, γ, can be expressed as

$$\gamma = \frac{\mathrm{Thp_m}}{\mathrm{Thp_n}} = \frac{\mathrm{P}_A(\mathrm{N1,N2,S,L})}{\mathrm{P}_B(\mathrm{N1,N2,S,L})} = \frac{\mathrm{P}'_A \sum_{i=0}^{S-1}\left(\prod_{j=2}^{N1}(S-i)\mathrm{P}'_A \times \prod_{k=1}^{N2}(L-i)\mathrm{P}'_B\right)}{\mathrm{P}'_B \sum_{i=0}^{S-1}\left(\prod_{j=1}^{N1}(S-i)\mathrm{P}'_A \times \prod_{k=2}^{N2}(L-i)\mathrm{P}'_B\right)} = \frac{\sum_{i=0}^{S-1}\left((S-i)^{N1-1} \times (L-i)^{N2}\right)}{\sum_{i=0}^{S-1}\left((S-i)^{N1} \times (L-i)^{N2-1}\right)} \quad (5)$$

According to the calculation formula, given the value of γ, N1 and N2, we can get a similar S.

4.2 Collision Problem

Similar to T2F, WT2F adopts two rounds contention to reduce collision probability. To be different with T2F, WT2F nodes pick subcarrier on their own ranges in Round 1 to achieve weighted contention. In Round 2, WT2F performs the same scheme that winners will pick subcarriers uniformly to contend channel. If more than one node picks the smallest subcarrier in Round 2, the collision will happen. According to T2F, the collision probability in WT2F is smaller than traditional 802.11. So the collision cannot be much effect on our analysis in Section 4.1.

4.3 Overhead in WT2F

According to T2F, the lack of synchronization in initiating subcarrier transmission emerges from: (1) two nodes may not observe the channel to become idle at the exact same time due to difference in propagation delays (t_{pd}), and (2) once all nodes begin subcarrier signaling, the signals arrive at any receiver with some stagger, caused by t_{pd} again. This total stagger introduces difficulty in "catching" all the active subcarriers at the same time – all the signals need to overlap at a receiver's antenna for at least one FFT window. WT2F copes with this problem by requiring the subcarrier signaling to occur for slightly longer duration than one OFDM symbol. This longer duration includes the maximum difference between propagation delays (= $2\mathrm{t}_{pd}$), time for the FFT computation, a hardware circuit delay, and guard factors. With $\mathrm{t}_{pd} = 1\mu s$ in WLANs, 64pt FFT taking $3.2\mu s$ at 20MHz, and circuit delay of $3\mu s$, two rounds of WT2F contention incurs $16.4\mu s$. This is appreciably smaller than time domain backoff, varying uniformly between $9\mu s$ to $135\mu s$ [4].

5 Performance Evaluation

In this section, we mainly demonstrate the effectiveness of the proportional fairness between HP and LP nodes via simulation. Let N1 and N2 denote the number of HP and LP nodes, respectively. Let S and L denote the maximum subcarrier number selected by HP and LP nodes, respectively. In simulation, by default, we set N1=10, N2=10, L =52, and S= 1~52 in Fig. 5 and set N1=1~50, N2=10, L =52, S=10 in Fig. 6. For each simulation result in each figure, we repeat the same experiment for 100000 times via different random seeds. For each theoretical result, we calculate the result by (6). In the following, we explain our observations in detail.

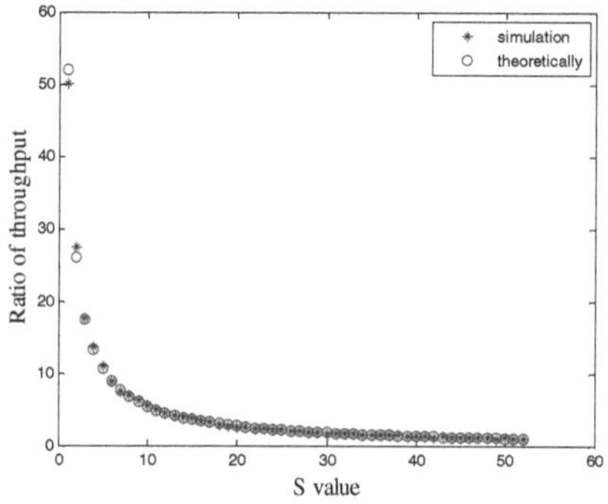

Fig. 5. Ratio of throughput between HP and LP nodes as S varies

Fig.5 plots the throughput ratio (γ) between HP and LP nodes when S = 1,2,⋯, 52. In this figure, X-axis is S value and Y-axis is γ . From this figure, we can see that 1) the theoretically values is almost the same as simulation values. 2) The ratio decreases (γ value reduces from 52 to 1) with the increasing of S value (S varies from 1 to 52). This is because that when S value varies from 1 to 52, the probability that HP nodes win the contention is becoming reduced. That means the proportional fairness can be achieved by setting different S value. On the other hand, determining the value of S, we can get the throughput ratio between LP and HP nodes. The ratio can be ranged from 1 to 52.

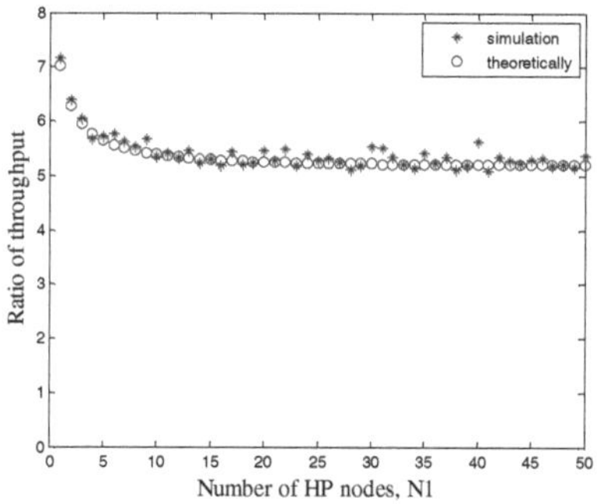

Fig. 6. Ratio of throughput between HP and LP nodes as N1 varies

Fig.6 plots the throughput ratio (γ) between HP and LP nodes when N1 varies from 1 to 50. From this figure, we have the following observations that 1) the theoretically values is much close to simulation values, 2) if N1 is less than 10, the throughput ratio is decreasing with the increasing of N1, 3) if N1 is more the 10, the throughput ratio is similar fixed. The reasons for 2) and 3) are that if the N1 is much less, each HP nodes can get more access probability, while when N1 is more, N1 increasing can do little effect on the throughput ratio.

6 Conclusion

In this paper, we design a weighted frequency contention based on T2F. In our design, nodes with different priorities select different subcarrier range so as to achieve different channel access opportunities. We then analyze the proportional fairness of our design. Finally, simulations verify that the proposed analysis is accurate.

Acknowledgment. This work is supported by the Macao Science and Technology Development Fund under Grant 081/2012/A3, Grant 013/2014/A1, Grant 018/2014/A1, and Grant. 096/2013/A3.

References

1. Heegard, C., Coffey, J., Gummadi, S., Murphy, P.A., Provencio, R., Rossin, E.J., Schrum, S., Shoemake, M.B.: High-Performance Wireless Ethernet. IEEE Comm. Magazine 39(11) (2001)
2. Sen, S., Choudhury, R.R., Nelakuditi, S.: Listen (on the frequency domain) before you talk. In: Workshop Hot Topics in Networks (HOTNETS), Monterey, CA, USA (2010)
3. Sen, S., Choudhury, R.R., Nelakuditi, S.: CSMA/CN: Carriersense multiple access with collision notification. IEEE/ACM Transactions on Networking 20(2), 544–556 (2012)
4. IEEE Standard: 802.11TM: Wireless LAN Medium Access Control(MAC) and Physical Layer (PHY) Specifications (2007)
5. Jardosh, A.P., Ramachandran, K.N., Almeroth, K.C., Belding-Royer, E.M.: Understanding Congestion in IEEE 802.11b Wireless Networks. In: Internet Measurment Conference, pp. 279–292 (2005)

Illustrating PNC Using Rectangular Pulses

Li Feng*

Faculty of Information Technology,
Macau University of Science and Technology,
Avenida Wei Long, Taipa, Macau, China
lfeng@must.edu.mo

Abstract. Physical-layer network coding (PNC) opens up a whole new research area because it innovatively proposes implementing upper-layer functions (i.e., network coding) at the physical layer. Study shows that PNC can achieve 100% throughput improvement compared with the traditional scheme. This paper aims at providing an entry-level tutorial that can help readers quickly and thoroughly understand the basic idea of PNC. In this paper, we first overview basic assumptions, bitwise XOR operation, network coding (NC), and then illustrate the decoding process and the performance of PNC using the example of rectangular pulses.

Keywords: PNC, rectangular pulses, decoding, performance.

1 Basic Assumptions

We consider a two-way relay channel (TWRC) in wireless networks, as shown in Figure 1. In TWRC, nodes 1 and 3 (i.e., N1 and N3) exchange information via the relay (i.e., N2) because they cannot hear each other.

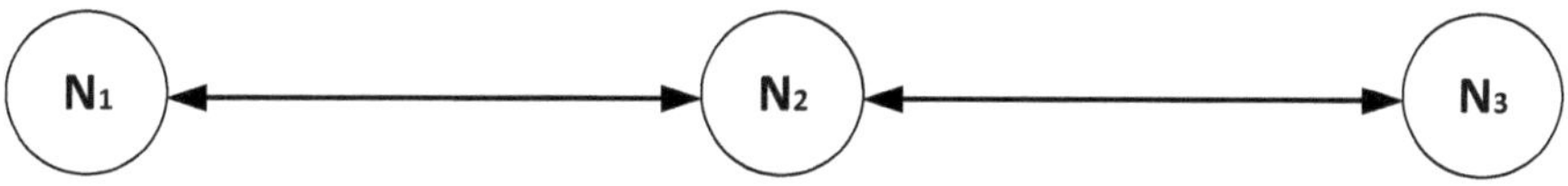

Fig. 1. Two-way relay channel

In TWRC, we assume that each node (i.e., N1, N2, and N3) only transmits two types of rectangular pulses, as shown in Figure 2:

- Pulse (a): a pulse of amplitude = 1 and duration = 1
- Pulse (b): a pulse of amplitude = -1 and duration = 1.

Pulse (a) and (b) are decoded as bit 1 and 0 at the receiver; conversely, if a node transmits 1 and 0, it will modulate them into pulse (a) and (b) at the physical layer, respectively.

---

* Corresponding author.

K.J. Kim and N. Wattanapongsakorn (eds.), *Mobile and Wireless Technology 2015*,
Lecture Notes in Electrical Engineering 310, DOI: 10.1007/978-3-662-47669-7_9

Time is divided into a series of slots with duration = 1. To simplify explanations, we have two assumptions:

- Assumption (a): Ideal channel conditions (i.e., no noises during communication)
- Assumption (b): Perfect synchronization (i.e., each node has the same clock and pulses are always transmitted at the beginning of each slot)

Later we will remove assumption (a) in Sections 4.2 and 4.3.

With assumptions (a) and (b), when N1 and N3 transmit a pulse (i.e., pulse (a) or (b)) simultaneously, N2 will received one of three superimposed pulses, as shown in Figure 3:

- Superimposed pulse (a): a pulse of amplitude = 2 and duration = 1
- Superimposed pulse (b) : a pulse of amplitude = 0 and duration = 1
- Superimposed pulse (c) : a pulse of amplitude = -2 and duration = 1

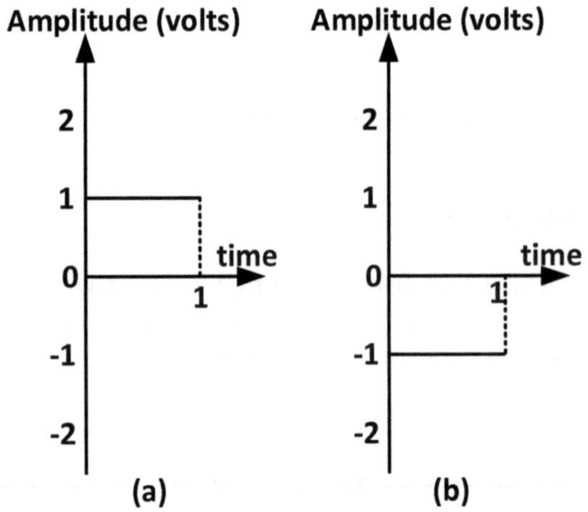

Fig. 2. Physical signals of duration 1: (a) rectangular pulse with amplitude = 1, and (b) rectangular pulse with amplitude = -1

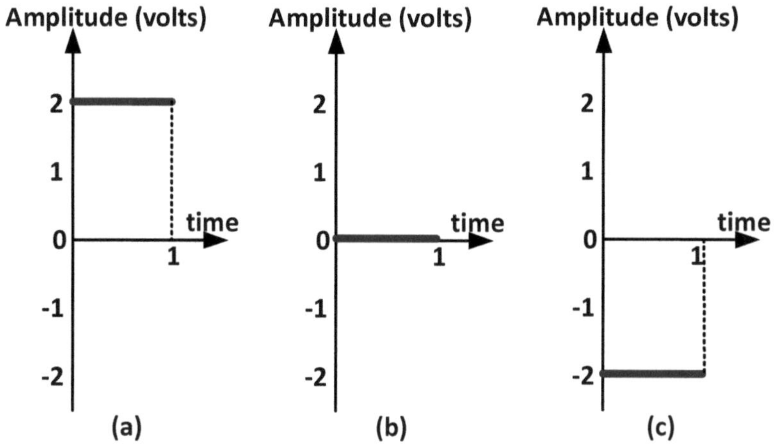

Fig. 3. Superimposed physical signals of duration 1: (a) rectangular pulse with amplitude = 2, (b) rectangular pulse with amplitude = 0, and (c) rectangular pulse with amplitude = -2

2 Bitwise XOR Operations

Let ⊕ to denote a bitwise XOR operator operating on binary bits, 0, 1. That is, 0⊕0=0, 1⊕1=0, 0⊕1=1, 1⊕0=1.

3 NC (Network Coding)

Let us consider NC for TWRC shown in Figure 4. Assume that N1 and N3 transmit bit 1 to each other. NC achieves this goal in three slots:

- In slot 1, N1 transmits pulse (a) and N2 decodes it to 1.
- In slot 2, N3 transmits pulse (a) and N2 decodes it to 1.
- In slot 3, N2 performs the bitwise XOR operation at the network layer, i.e., 1⊕1=0, and therefore broadcasts pulse (b) at the physical layer, and then
  - N1 will (i) decode pulse (b) to 0 and (ii) perform the bitwise XOR operation between the received 0 and its own 1 (i.e., 0⊕1=1). Then N1 knows that N3 transmits bit 1 to itself.
  - N3 will perform the same operation as N1.

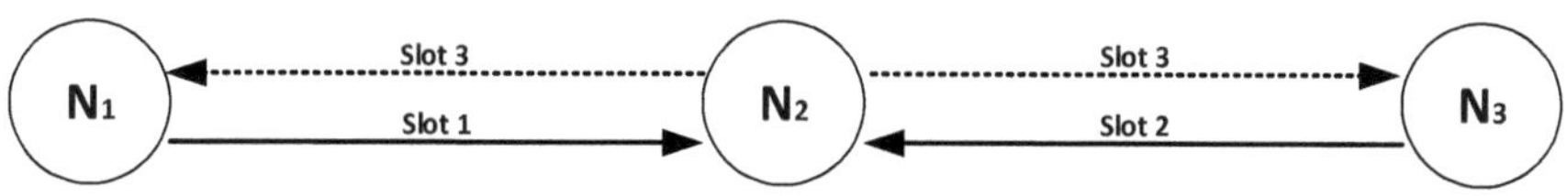

Fig. 4. NC for two-way relay channel

4 PNC (Physical-Layer Network Coding)

4.1 Overview of PNC

Let us consider PNC for TWRC shown in Figure 5. Assume that N1 and N3 transmit bit1 to each other. PNC achieves this goal in two slots:

- In slot 1, N1 and N3 transmits pulse (a) simultaneously; N2 will receive a superimposed signal. Under ideal channel conditions, the received signal is the superimposed pulse (a) as shown in Figure 3.
- In slot 2, N2 transforms superimposed pulse (a) (in Figure 3) to pulse (b) (in Figure 2), which corresponds to N2's XOR result, 0, in NC of Section 3. This transformation process is called a PNC mapping. Then N2 broadcasts pulse (b). The subsequent process is completely same to that in NC.

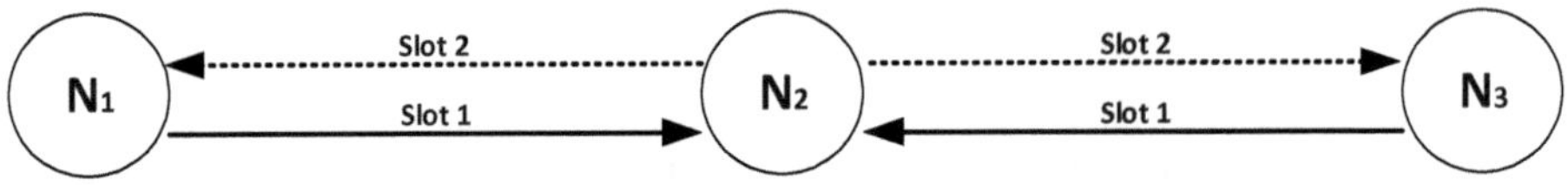

Fig. 5. PNC for two-way relay channel

The main difference between NC and PNC lies that (i) N2 in PNC will receive a superimposed signal, (ii) N2 extracts useful information from the received superimposed signal, and then transforms the extracted information into a network-coded form (i.e., NC's XOR results here). After this, we can leverage NC to perform the subsequent procedure.

We define the PNC decoding process to be the process that the relay handles the reconceived superimposed signals and transforms them into a network-coded form.

4.2 PNC Decoding Process

In this section, we present the PNC decoding process [3] under non-ideal channel condition, but we still assume perfect synchronization. Note that analog network coding in [5] has been proposed to remove the requirement of synchronization.

Figure 6 plots the multiple access diagram for PNC in TWRC. In this diagram, we assume that N1 modulates bit s1∈{0,1} into a physical signal a1, and at the same time N2 modulates bit s2∈{0,1} into a physical signal a2. Let w denote the noise. Then N2 will received a superimposed signal with noise, i.e., r = a1+a3+w. After that, N2 begins decoding the superimposed signals.

PNC decoding process is composed of two sub-process: signal detection and PNC mapping (see the blue-dashed rectangular box in Figure 6). The purpose of PNC decoding is to decode a1+a3+w into the network-coded form s1⊕s3, which corresponds to the XOR result in N2 of NC.

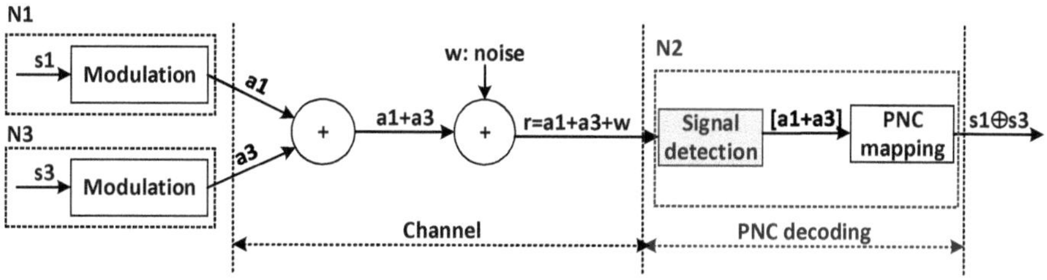

Fig. 6. Multiple access diagram for PNC in TWRC

Signal Detection Sub-process: This sub-process is to estimate the useful superimposed signal, [a1+a3], from the combined signal with noises, a1+a3+w. Under ideal channel conditions, i.e., w=0, we have r=a1+a3=[a1+a3]. [a1+a3] must be one of three superimposed pulses in Figure 3. Here, traditional methods can be used for superimposed signal detection: hard decision and soft decision, which identify a signal according to the given threshold and probability, respectively.

We now explain the hard decision, with the help of Figure 7. In this example, we set two thresholds, r1 and r2 (whose expressions can be found from Eq. (8) in [1] [2]), to identify one from the three superimposed pulses in Figure 3 whose amplitudes are -2, 0, 2. For example, we can declare that the received signal is the superimposed pulse (a), (b), and (c) if its amplitude is within [r2,), (r1, r2), and (∞, r1], respectively.

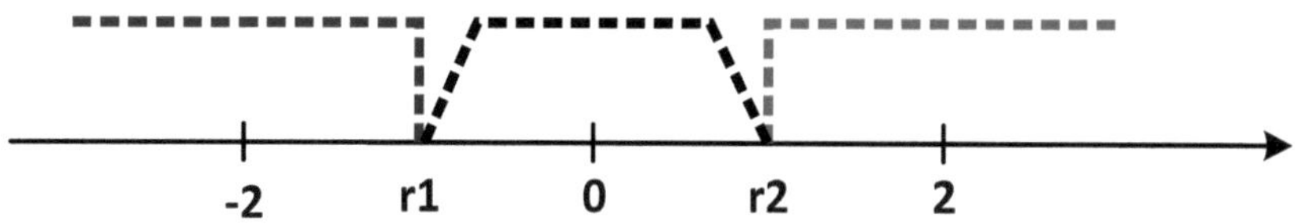

Fig. 7. An example for hard decision

PNC Mapping Sub-process: PNC mapping refers to the process of mapping the superimposed electromagnetic waves under ideal channel conditions to some network-coded form for forwarding by the relay. The complete PNC mapping can be found in Table I in [1] [2].

In Figure 6, we will map the estimated signal, [a1+a3], into the network-coded form s1⊕s3. If [a1+a3] is the superimposed pulse (a) in Figure 3, then it will be mapped to pulse (b) in Figure 2, as explained in Section 4.1.

4.3 PNC Performance

In this section, with the help of Figure 8 (which is from Fig. 14 of [2]), we show that PNC is a very promising and practical technology.

Figure 8 plots the channel capacity of TWRC as the SNR varies. From this figure, we can see that when SNR ≥ 5dB, PNC can achieve 100% and 50% capacity increases compared with traditional transmission and NC. In wireless LAN, the typical SNR value is larger than 10dB from [4]. This manifests that the 100% and 50% capacity increment in PNC is practicable.

What's more, Figure 8 shows that PNC can do better than traditional transmission and NC. For example, when -5≤ SNR <5dB, the lower bound capacity of PNC is better than the capacity of the latter; when SNR < -5dB, the upper bound capacity of PNC is better than the capacity of the latter.

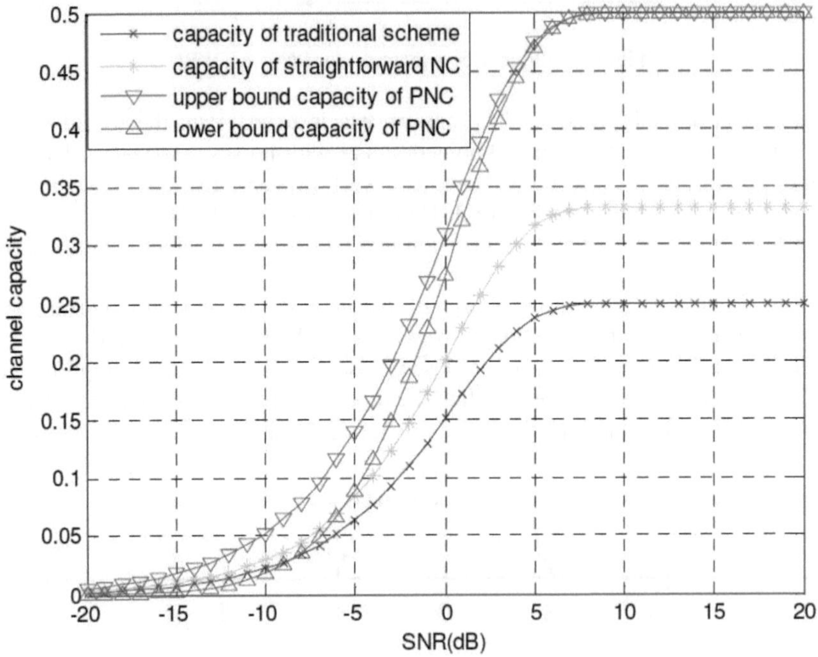

Fig. 8. Channel capacity comparisons of traditional scheme, NC, and PNC

5 Conclusion

PNC is a typical example that exploits the physical-layer interference to improve the network performance. Its basic idea is to leverages the superimposition feature of electromagnetic (EM) waves from multiple nodes (which is generally considered harmful in conventional wireless networks) to implement the network coding operation. This paper simply and clearly illustrates the basic ideal of PNC by the example of rectangular pulses. It is a very good tutorial that can help readers understand PNC quickly.

Acknowledgment. This work is supported by the Macao Science and Technology Development Fund under Grant 013/2014/A1 and Grant 081/2012/A3.

References

1. Zhang, S., Liew, S.C., Lam, P.P.: Hot topic: physical-layer network coding. In: MOBICOM 2006, pp. 358–365 (2006)
2. Zhang, S., Liew, S.C., Lam, P.P.: Physical layer network coding, http://arxiv.org/ftp/arxiv/papers/0704/0704.2475.pdf
3. Zhang, S., Liew, S.C.: Channel coding and decoding in a relay system operated with physical-layer network coding. IEEE Journal on Selected Areas in Communications 27(5), 788–796 (2009)
4. http://www.wireless-nets.com/resources/tutorials/define_SNR_values.html
5. Katti, S., Gollakota, S.S., Katabi, D.: Embracing Wireless Interference: Analog network Coding. In: ACM SIGCOMM 2007, pp. 397–408 (August 2007)

Application of Wireless Personal Area Networks to a Ship Area Piconet

Dong-Keun Jeon[1], Yeonwoo Lee[2,*], and Kyung-Ho Kim[2]

[1] Dept. of Mechatronics, School of Mechanical Engineering, Incheon National University, Incheon, Korea
[2] Dept. of Information and Communications, College of Engineering, Mokpo National University, Chonnam, Korea
dkjeon@incheon.ac.kr, {ylee,khkim}@mokpo.ac.kr

Abstract. An integrated ship area network (SAN) architecture of the international standard is mainly based on wired instrument network connections and shipboard control networks. However, for a high-speed data transmission support to onboard instruments or instrument-to-instrument, we apply IEEE 802.15.3 wireless personal area networks (WPAN) to a high-cost valued ship area network. Thus, this paper consider two IEEE 802.15.3 WPAN protocols for a ship area piconet employing a wireless gateway concept. With this, it is shown that the WPAN gateway application can deliver a high quality transmission with energy efficiency and a higher data throughput performance. Moreover, it can be seen that a realistic transmission parameter choice for two WPAN protocols is essential to guarantee both less energy consumption and high data throughput.

Keywords: ship are network, instrument network, WPAN, wireless gateway, piconet.

1 Introduction

The ship area network (SAN) architecture of the international standard has many functionalities of dedicated Ethernet based wired connection, controller area network connection and instrument network connection, etc. These are standardized by the international maritime organization (IMO) or the international electro-technical commission (IEC) [1-2]. Navigational instruments are normally interconnected with the well-established IEC 61162-1 or IEC 61162-2 standards. These networks deliver operations such as control of shipboard systems and management of crucial information for safety and navigation. These operations are performed in many parts of the vessel; e.g., from each part of instruments area to the bridge, to the administrative personnel, and even off of the ship to the owner's office. However, such a conventional wired network hierarchy based on IEC61162-3 and NMEA2000 standards can only support data rate of form 4.8 kbps up to 250 kbps. By this reason, this paper focuses on applying IEEE 802.15.3 WPAN standards to the SAN

---

* Corresponding author.

K.J. Kim and N. Wattanapongsakorn (eds.), *Mobile and Wireless Technology 2015*,
Lecture Notes in Electrical Engineering 310, DOI: 10.1007/978-3-662-47669-7_10

architecture. For a near future high-value added ship, to satisfy the need for various and high-speed data services is essential. Moreover, such a conventional control area network cannot satisfy an increasing need for large amount of data transmission requirement on board between instruments.

Therefore, for supporting various and high-speed data service, this paper consider two IEEE 802.15.3 WPAN standard protocols with adopting a wireless gateway concept, which was introduced in our previous research work [3]. A WAPN pcionet coordinator (PNC) plays a wireless gateway in a ship area piconet to coordinate transmission between instruments. With this concept, IEEE 802.15.3 WPAN is capable of providing reliable transmission and flexible network deployment with high quality of services. However, this have considered only a TDMA WPAN protocol application. Thus, this paper considers two WPAN standards and compares the performance of these two protocols with a wireless gateway applied to a ship area piconet environment. That is, WPAN gateway can be operated with either TDMA based MAC protocol or CSMA/CA based MAC protocol, which is so as to minimize energy consumption of instruments as well as to increase data throughput. Moreover, this paper evaluates and compares the performance effect of transmission parameter selection of different MAC protocols. Section 2 in this paper describes the IEEE 802.15.3 WPAN based ship area piconet with a WPAN gateway. Section 3 shows the simulation results and discussion of two WPAN standards applications for a ship area piconet, followed by the conclusion.

2 IEEE 802.15.3 WPAN Based Ship Area Piconet with a WPAN Gateway

2.1 Structure of IEEE 802.15.3 WPAN Based Ship Area Piconet

The conventional shipboard data architecture by the IEC standard is the Maritime Information technology Standard (MiTS) project [4]. NMEA 2000 is IEC-accepted as a standard IEC 61162-3 on IEC technical committee 80, working group 6 (digital interfaces), and successfully develops another LAN (Local Area Network) standard for ships where LAN requires high capacity and also high security (IEC 61162-4). This was developed as an integrated ship control (ISC) protocol, which could integrate instrument networks (NMEA 2000) on the gateway with industrial data network in the automation system. This system will allow up to about 50 nodes to share a common bus at 250 kbps. Messages vary from 8 bytes up to about 1,000, but at an overhead of 50%. Thus, effective maximum throughput on the bus is about 125 kbps. Among four layers such as instrument, process, system, and administrative, at the instrument layer, instruments and devices are connected to a wired gateway. However, this cannot support sufficient data throughput caused by various instruments and devices, etc.

Thus, this paper applies an architecture of IEEE 802.15.3 WPAN ship area piconet model with WPAN gateways. This conceptual model is demonstrated in Fig. 1 where IEEE 802.15.3 WPAN constructs a ship area piconet and it plays an important role in coordinating WPAN devices' communications. This WPAN gateway connects wireless data transmission between WPAN devices (DEV) instruments which are connected to various instruments such as throttle control, engine computer, transmission control, depth sounder, compass, fuel control, and actuators/sensors. These instruments are components of a ship board control network. As shown in Fig. 1, each WPAN DEV is a wireless transceiver and data collector connected to various navigational instruments or actuators/sensors. With this model, a WPAN ship area piconet is sufficiently capable of supporting high-speed data transmission and providing a degree of freedom in deploying network with reliable high quality and energy efficiency. Thus, this paper consider IEEE 802.15.3 WPAN ship area piconet with either CSMA/CA based or TDMA based MAC protocol. It should be noted that IEEE 802.15.3 WPAN standards were not discussed in standardization any more. However, since the traffic size and the coverage of ship area piconet are quite adequate to the IEEE 802.15.3 WPAN standards, we consider these standards as a piconet model for realizing a ship are piconet application.

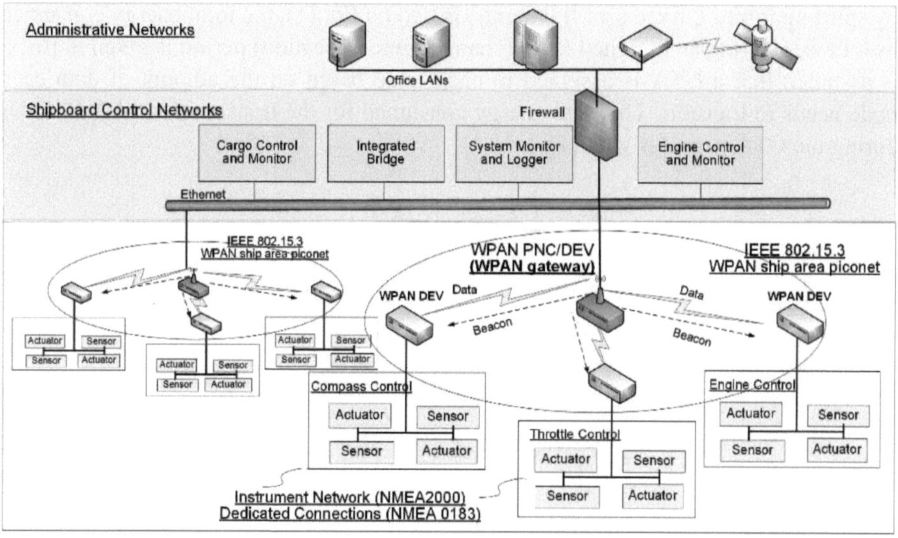

Fig. 1. The considered IEEE 802.15.3 WPAN ship area piconet with WPAN gateways

2.2 Transmission Parameter Selections

In this paper, it is assumed that IEEE802.15.3 WPAN system supports 3 data rates, i.e., 11, 22 and 33 Mbps with QPSK-TCM, DQPSK-TCM and 16QAM-TCM, respectively [5].

TDMA Based MAC Protocol. For TDMA based MAC protocol, the energy required to transmit a frame using fragment size L, data rate R and transmission power P_t, is given by $E_{frame}(L,R,P_t)=T_{frame}(L,R)P_{tx}(P_t)$ where T_{frame} is the transmission time of the frame, P_{tx} is the total power required to transmit the frame considering the efficiency of the power amplifier and modulation circuit. When transmitting an entire MAC service data unit (MSDU) with a size of L_{MSDU}, a device must transmit $\lceil L_{MSDU}/L \rceil$ frames. Therefore, the energy required to transmit an entire MSDU without consideration on transmission errors is obtained as follows:

$$\begin{aligned} &E_{MSDU}(L,R,P_t) \\ &= \left\lfloor \frac{L_{MSDU}}{L} \right\rfloor \cdot E_{frame}(L,R,P_t) + E_{frame}(L_{remain},R,P_t) \\ &+ \left\lceil \frac{L_{MSDU}}{L} \right\rceil \cdot (E_{SIFS} + E_{ACK}) \end{aligned} \quad . \tag{1}$$

where L_{remain} is the length of the remaining data after the MSDU has been fragmented by splitting it into L fixed size fragments. Based on Eq. (1), the total energy consumed by a device during an assigned CTA (channel time allocation) period is given in [6]. It is assumed that a CTA is assigned to each node based on the amount of data each node needs to transmit. The total energy consumed for the transmission of K MSDUs during the CTA period is given by

$$E_{CTA} = \sum_{k=1}^{K} E_{MSDU,k}(L,R,P_t) \quad . \tag{2}$$

where

$$\begin{aligned} E_{MSDU,k}(L,R,P_t) = \Bigg(&\left\lfloor \frac{L_{MSDU,k}}{L} \right\rfloor \cdot E_{frame}(L,R,P_t) + E_{frame}(L_{remain},R,P_t) \\ &+ \left\lceil \frac{L_{MSDU,k}}{L} \right\rceil \{(E_{SIFS} + E_{ACK})(1 - FER) + E_{RIFS} \cdot FER\} \Bigg) \cdot \sum_{i=0}^{q} FER^i \end{aligned} \tag{3}$$

as presented in [6].

CSMA/CA Based MAC Protocol. When applying CSMA/CA based MAC protocol, the energy required to transmit a frame using fragment size L, data rate R, and transmission power P_t is given by $E_{frame}(L,R,P_t) = E_{rx} + E_{tx}(L,R,P_t) = E_{CS} + E_{ACK} + E_{tx}(L,R,P_t)$ where E_{rx} is the receiving energy corresponding to the energy required for carrier sensing (E_{CS}) and ACK reception (E_{ACK}), and the data transmission energy (E_{tx}). With some analytic calculation of energy for carrier sensing by [7], the expected energy required to transmit an entire MSDU (MAC service data unit) is obtained as follows:

$$E_{MSDU}(L,R,P_t) = \left\lfloor \frac{L_{MSDU}}{L} \right\rfloor \cdot E_{frame}(L,R,P_t) + E_{frame}(L_{remain},R,P_t) \quad . \tag{4}$$

Transmission Parameter Selection. For minimizing the energy required to transmit a MSDU as in Eq. (2) and Eq. (4), the WPAN piconet should find the transmission parameter combination of [L, R, P_t]. This procedure is performed based on measurement information of received signal strength indicator (RSSI) and link quality indicator (LQI). With such transmission parameter combination, the WPAN ship area piconet can choose the optimal transmission parameter sets. However, this cannot always guarantee practical solution to such network. It was concluded that the selection of a specific transmission parameter was most time better than the updated parameter combination set at every time allocation, from the previous our research result in [3]. Thus, this paper evaluates the performance comparison of WPAN ship area piconet based on either CSMA/CA or TDMA MAC protocol with a specific transmission parameter with periodically updated other parameters. Moreover, it is noted that the path loss is a most important parameter to decide the overall performance of the WPAN ship area piconet, whether the optimal transmission parameter combination set is used or not. Because the path loss are very dependent on material of the ship, in our work, three cases such as steel, FRF (fiber-reinforced plastic) and wood are considered by setting path loss value to 60, 55 and 50dB, respectively.

3 Simulation Results and Discussion

3.1 Simulation Parameters

As a simulation platform, we use Matlab simulation tool for the performance evaluation of the IEEE 802.15.3 WPAN ship area piconet model with wireless gateways concept. Most parameters assumed in this simulation are based on IEEE 802.15.3 physical parameter and energy parameters [5], [6], [7]. We assumed that for setting up ship area piconets, two WPAN PNC/DEVs act as WPAN gateways and 10 WPAN devices are connected to shipboard instruments. These devices are randomly distributed. It is assumed that a fragment size is fixed to 256 bytes and the maximum transmission power is between 0 and 8dBm with step size of 2 dBm. The other parameters for either MAC protocol are given in Table 1. As shown in Table 1, P_c is the power consumed in modulation and demodulation circuit and P_r is the power consumed at the receiving front end. These values are assumed as $P_{tx}=P_c+P_t/\eta$ and $P_{rx}=P_c+P_r$ where η represents the efficiency of the power amplifier (set to 0.1).

Without realistic channel measurement of real ship environment, we assumed that the path loss of ship environment would be higher than that of terrestrial environment and thus, in this simulation, three path loss cases such as 50dB, 55dB and 60dB are considered as small wood ship, FRF ship, and steel ship, respectively. This is to consider the effect of material of the ship construction in our simulation.

Table 1. IEEE 802.15.3 based HR-WPN physical layer simulation parameters [5], [6], [7]

Parameter	Value	
	CSMA/CA MAC protocol	TDMA MAC protocol
Simulation time	60ms (10 CTA) * 100 times	
MSDU size	3,000 bytes	
Fragment size	256, 512, 1024 bytes	
Data rate	11, 22, 33Mbps	
Target FER	0.2	
T_{ACK}	22.54 μs	22.54 μs
T_{SIFS}	10 μs	10 μs
T_{RIFS}	27.27 μs	27.27 μs
T_{BIFS}	17.27 μs	-
P_c , P_r	400 mW, 50 mW	
Path loss	50 dB, 55 dB, 60 dB	

3.2 Simulation Results

Figure 2 shows the average energy consumption comparison of all connected devices to IEEE 802.15.3 WPAN based ship area piconet. As shown in Fig. 2, it is obviously shown that for the fixed transmission power, as noise power level increases above -60dBm, the connected device consumes more energy than that of below -60dBm. The similar result can be observed when the path loss increases. It is noted that fixing the transmission power means the network updates other two parameters, i.e., data rate and fragment size. Since in case of high noise level where the target FER cannot be achieved, the devices are forced to increase transmission power and decrease the fragment size. Since it deteriorates the communication channel status like the noise level, the similar result can be seen when the path loss increases.

Thus, the devices with the fixed transmission power repeatedly try to succeed the connection and this process consumes more energy than in case of lower noise level. This observation can be clearly seen that when comparing the energy consumption with the fixed data rate case, both MAC protocol with the fixed transmission power require less energy consumption below -60dBm. The other observation in Fig. 2 is that the ship area piconet with CSMA/CA MAC protocol requires somewhat less energy than that with the TDMA MAC protocol in case of high noise level.

The average data throughput performance comparison is shown in Fig. 3. As shown in Fig. 3, it can be seen that in case of low noise level where the target FER can be easily guaranteed, the higher data throughput with either higher transmission power or TDMA protocol can be achieved. However, as noise power level increases above -60dBm, the connected device cannot achieve higher data throughput because of lack of transmission power to communicate with other devices. On the other hand, the device with the fixed data rate parameter can successfully communicate with other devices even above -60dBm of noise level. It is noted that the device with higher transmission power (i.e. 8dBm) can overcome severe channel condition caused by either high noise level or high path loss. Moreover, even when the noise level is higher than -60dBm, it can outperform that with the fixed data rate selection.

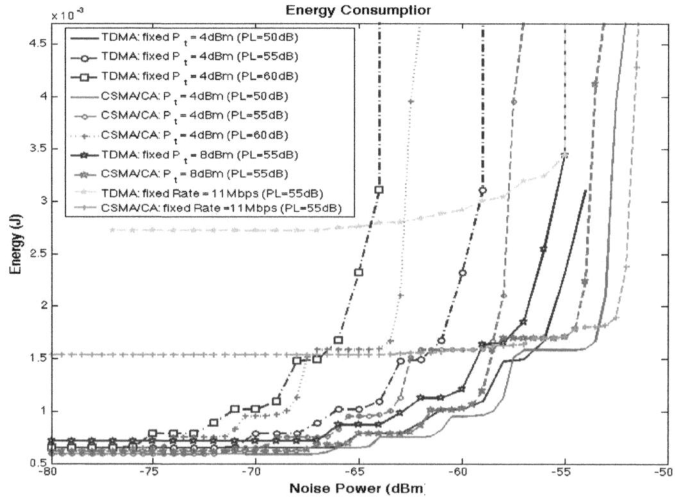

Fig. 2. Average energy consumption comparison of two WPAN protocols as a function of the variation of noise power, transmission parameter (P_t, R) and path loss

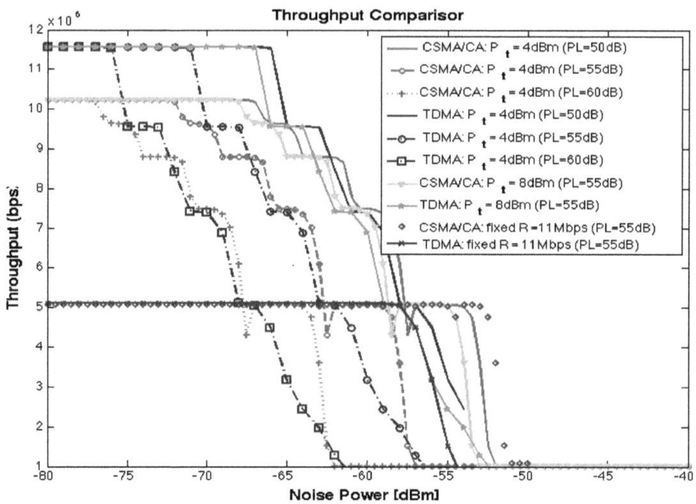

Fig. 3. Data throughput comparison of two WPAN protocols applied to a ship area piconet as a function of the variation of noise power, transmission parameter (P_t, R) and path loss

From both simulation results, it can be concluded that in case of low noise level and path loss, the devices with the fixed transmission power is a more reasonable choice than that with the fixed data rate. For higher noise level or path loss, to increase the transmission power is a better option. Moreover, it is demonstrated that IEEE 802.15.3 WPAN based ship area piconet with wireless gateways can be a reasonable option for satisfying the need for various data services within a vessel is essential for a high-value added vessel by virtue of energy efficiency and system deployment cost.

4 Conclusion

In this paper, we evaluated the energy consumption and data throughput performance of IEEE 802.15.3 WPAN based ship area piconet with a wireless gateway concept. It was confirmed that the WPAN gateway application with the proper link parameters selection could provide less energy consumption and higher achievable data throughput. From simulation results, it was concluded that in case of low noise level and path loss, the devices with the fixed transmission power is a more reasonable choice than that with the fixed data rate. For higher noise level or path loss, to increase the transmission power is a better option. Thus, although IEEE 802.15.3 standards are obsolete, IEEE 802.15.3 WPAN based ship area piconet with adaptively selected transmission parameter can be successfully applicable to a high-value added vessel.

Acknowledgments. This research was supported by Basic Research Program through the National Research Foundation of Korea (NRF) funded by the Ministry of Education, Science and Technology (No. 2013R1A1A2008722).

References

1. Maritime Information T Standards (MiTS), http://www.mits-forum.org/architecture.html
2. Krile, S., Kezić, D., Dimc, F.: NMEA Communication Standard for Shipboard Data Architecture. Our Sea, International Journal of Maritime Science & Technology 60(3), 68–81 (2013)
3. Jeon, D., Lee, Y.: Performance Evaluation of a WiMedia based Wireless Bridge using Relay Cooperative Transmission. Advanced Science and Technology Letters 60, 1–5 (2014)
4. Rødseth, Ø.J., Haaland, E.: MiTS-An Open Standard for Integrated Ship Control. In: Proceedings of ICMES 1993, Hamburg (September 1993)
5. IEEE 802.15.3, Part 15.3: Wireless MAC and Physical Layer WPAN Standard. IEEE (2003)
6. Joo, Y., Lee, Y.: Energy-Efficient Transmission Scheme for WPANs with a TDMA-based Contention-Free Access Protocol. IEICE Transactions on Communications E91-B(2), 609–612 (2008)
7. Joo, Y., Lee, Y.: QoS Guaranteed and Energy Efficient Transmission Scheme for Wireless Personal Area Networks. IEICE Transactions on Communications E91-B(6), 1980–1988 (2008)

Design and Development of a Wireless Sensor Network Framework for Water Quality Remote Monitoring

Francis Jerome Tiausas*, Maria Leonora Guico, Jose Claro Monje, and Carlos Oppus

Electronics, Computer, and Communications Engineering Department
Ateneo de Manila University
Loyola Heights, Quezon City, Philippines
francis.tiaus@gmail.com,
{mguico,jcmonje,coppus}@ateneo.edu

Abstract. This study involves the design and development of a wireless sensor network (WSN) that integrates several sensing modules into a fully-functional system. The overall system is composed of a remote server, a controller node, and several sensing modules. The controller node is implemented using an Android mobile phone with Bluetooth and 3G capabilities. Bluetooth is used to communicate with the various sensing modules; while 3G is used to relay data to the remote server. The sensing modules utilize an Arduino Mega 2560 (with the sensor circuits) and a Bluetooth shield. Test results show that this framework is a viable design for WSN systems and can be used for remote installations that can be continuously upgraded over time.

1 Introduction

The importance of a stable and usable water supply is of great concern to modern day society. It is utilized in a wide variety of applications, ranging from industrial (e.g., cooling) to agricultural (e.g., irrigation).[1] Its most important use is to provide drinkable water to the population. It is imperative that water quality from these sources remain consistently within the standards identified as safe for human consumption.[2]

One important way of ensuring that water quality remains consistent and up to standards is regular scientific monitoring. The most widely-used approach for assessing water quality is by conducting field work and sampling in the areas of concern.[3] Though the results are reliable and accurate, it is also costly and requires a lot of effort in planning and execution. Another approach is to use automated loggers which take measurements at specific intervals and logs these on a storage medium. This, however, still requires the storage medium to be obtained manually before data can be analyzed [4], which takes time if the area is remote. The more modern approach is to deploy wireless sensor networks (WSNs) to remotely monitor such areas.

WSNs are sensor systems that can be deployed at remote sites. They can be made to autonomously perform measurements and send the results back to the user in real

* Corresponding author.

K.J. Kim and N. Wattanapongsakorn (eds.), *Mobile and Wireless Technology 2015*,
Lecture Notes in Electrical Engineering 310, DOI: 10.1007/978-3-662-47669-7_11

time through wireless link. They are usually composed of an array of sensor nodes. Data from these sensor nodes can be sent back through the sensor network, usually passing through a base station, to a remote server.[5] WSNs are ideal in environmental monitoring applications due to their flexibility, potential for autonomous operation, and ability to send data back from remote locations in real-time [4], which allows for the collection of high-resolution environmental data. [6]

WSNs provide the advantages of being low cost, low power, scalable, and capable of communicating data wirelessly.[7] Recently, WSNs have also started using commercial GSM service provider network to enable easy transmission of the real-time data collected from the sensors to a remote database server.[8]

This work deals with the challenges involved in the design and development of a WSN to integrate several sensing modules – each being developed in parallel by different teams from different organizations – into one fully-functional system. This potentially means having multiple subsystems with each having their own communication protocols for transmitting data, and receiving commands. Such a system would likely need a complex way of managing both the flow and format of data from such a wide variety of modules, and would be very inflexible when dealing with changes such as a modification in data format or the addition of an entirely new module.

This study proposes a design for a WSN framework that addresses these issues and implements it in a fully-functional water quality monitoring system.

2 WSN Framework Overview and Design

The overall system, shown in Fig. 1, is composed of three or more distinct entities as: the remote server, the controller node, and the sensing modules.

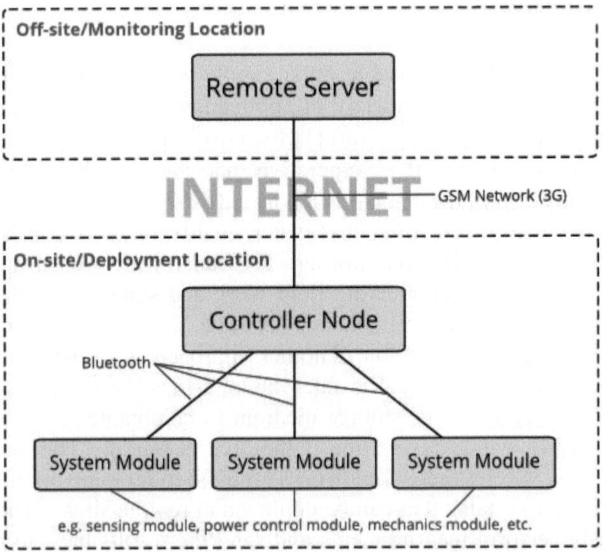

Fig. 1. Basic Design for the Wireless Sensor Network (WSN) System

The remote server functions both as a repository for data from the sensing modules. It also serves as a reference for metadata information that the controller node needs in order to operate the sensing modules connected to it.

The controller node functions as a local coordinator that manages one or more sensing modules. It transmits commands at specific times or when particular conditions are met, and relays information to the remote server.

The sensing modules are the system's agents which interact directly with, or gather information from the environment. To some degree, these modules are able to inform the controller of their status. As an example, in the system described in Fig. 1, a power module can monitor the amount of power left in an internal power source and inform the controller node the next time it is polled for its status. Similarly, a mechanics module can check if the mechanism for submerging the sensors in water is functioning correctly by checking a water level sensor attached to the sensor array.

These modules then have a simple and standard way of communicating with the controller node. The adopted communication configuration was a simple plain text Command-Response model over a wireless transmission medium – the controller node sends a command to a corresponding module, and the corresponding module performs the task and returns a response to the controller node. By standardizing the communication protocol, the integrating new modules become much easier.

3 System Implementation

The particular set of components used for the implementation of the design was chosen to be low cost but with enough processing power to meet the projected demands of the final system. Table 1 shows a list of these components and their respective roles in the system.

Table 1. WSN System Components and their corresponding roles

System Role	*Hardware Components*
Remote Server	---
Controller Node	Android Mobile Phone (Lenovo S820) - Bluetooth capability (built-in) - GSM/3G capability (built-in)
Water Quality Sensing Module	Arduino Mega 2560 E-Gizmo Bluetooth Shield Atlas Scientific pH sensor probe and circuit Atlas Scientific Dissolved Oxygen sensor probe and circuit Atlas Scientific Conductivity sensor probe and circuit
Dummy Sensing Module	Arduino Mega 2560 E-Gizmo Bluetooth Shield

The controller node is implemented with an Android mobile phone with Bluetooth and 3G communication capabilities. Bluetooth is used to communicate with the various modules associated with this controller node, while 3G is used for relaying information to the server. The phone also has a built-in camera used in conjunction with a chemical detection module. The controller node's entire logic is implemented through an Android Java application on the mobile phone.

The primary advantages of using an Android mobile phone as the controller node are (1) the necessary communication features needed for both the remote server and system modules; (2) having the minimum processing power needed to coordinate those operations; (3) being highly programmable (via Android applications); (4) relatively inexpensive, and (5) having other useful built-in features such as GPS, cameras, and accelerometers which can potentially be of good use to the researcher.

The system's modules utilize a combination of an Arduino microcontroller board with Bluetooth capability. In the particular case of the water quality sensing module used for this project, an Arduino Mega 2560 [9] was combined with a set of Atlas Scientific sensor circuits namely pH, dissolved oxygen, and electrical conductivity [10], and a Bluetooth shield. The system module logic is written using an Arduino variant of the C programming language.

4 Controller Node Logic

The basic execution logic of the controller node is shown in Fig. 2. Upon startup, the controller begins by reading through a configuration file saved on the Android mobile phone. The configuration file contains information on the how to communicate with the different modules and the proper sequence in which they should be operated. The Initialization step terminates after a certain number of failed attempts, otherwise the flow proceeds.

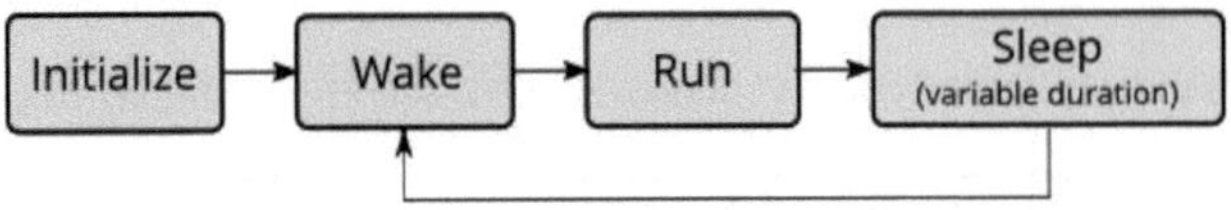

Fig. 2. Controller Node Logic Flowchart

During the Wake step, the controller node first checks if a new version of the configuration file is available on the remote server. If the local configuration file is outdated or missing, the controller then attempts to download a configuration file from the remote server to ensure that it is always up-to-date and valid. Afterwards, the controller node attempts to load the available configuration file.

During the Run step, the controller node sends the respective commands to each module in the sequence listed on the configuration file. The controller node then checks whether the received responses indicate success or failure. The commands are sent through Bluetooth in plain human-readable ASCII.

Once done with the entire Run step, the controller node enters the Sleep step, deactivating active connections with the modules to conserve power for a given period of time indicated on the configuration file. The controller node re-enters the Wake step upon finishing the Sleep step, or when a particular condition is achieved.

5 System Module Logic

All modules follow the same basic logic for communicating with the controller node – differing only in the processing step for commands and module-specific recurring tasks which fall outside or in between communication steps. The basic logic followed by system modules is shown in Fig. 3.

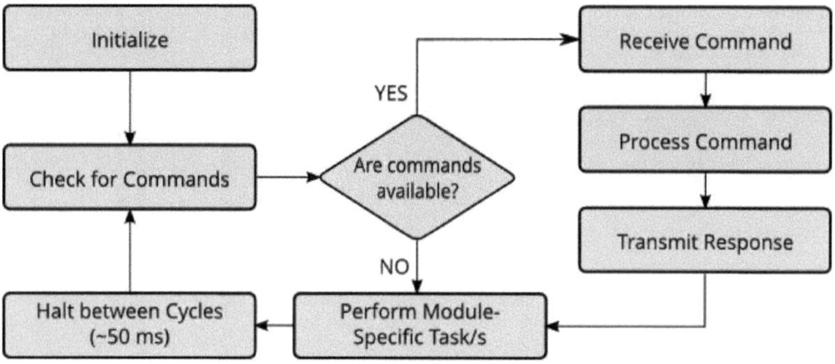

Fig. 3. System Module Logic Flowchart

At the start of the cycle, the module attempts to receive a command string from the controller node and store it in a buffer for processing. Afterwards, the module attempts to process the command and perform the task associated with it. The result of the task is then stored in another buffer for transmission. If successful, the module transmits the end result back to the controller node as a response. Otherwise, a failure message is transmitted back as a response.

6 Results and Analysis

6.1 WSN Prototype Equipment

In order to test the viability of the system framework, a prototype WSN system was constructed. The basic prototype consisted of a controller node, and a water quality sensing module as shown in Fig. 4. Though the system is already fully-functional in this iteration, a dummy sensing module was also developed in order to simulate the conditions of being able to operate different modules on-the-fly and having to operate more than one module at a time. This dummy module simply responds with randomized but appropriately bounded water quality data when given a READ command by the controller node. The water quality data transmitted by the prototype WSN system is sent to a remote server. The extended prototype which includes the dummy sensing module is shown in Fig. 4 as well.

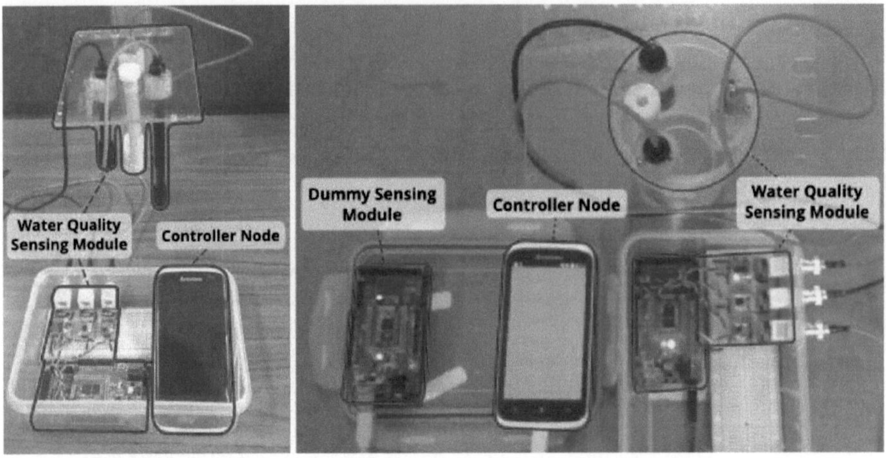

Fig. 4. WSN System Basic Prototype (left) and Extended Protoype (right)

The controller node was implemented using an Android mobile phone. Likewise, both sensing modules were implemented using Arduino microcontroller boards with Bluetooth shields. The actual water quality sensing module was further equipped with a pH sensor, a dissolved oxygen sensor, and an electrical conductivity sensor.

6.2 Additional Preparations

Since the mechanism for retrieving configuration files from the remote server was not yet in place during the last field test, a configuration file was directly created and stored in the controller node's memory. This skips the retrieval step entirely, but retains the configuration file loading step which would still give the controller node the required information to operate the sensing modules, and the correct order in which to transmit them.

A rudimentary Android application GUI was also developed in order to be able to see the system's results occurring in real-time. This basically consisted of a start button, a stop button, and a list to show the data received from the sensing modules.

6.3 Test Results and Findings

Initial testing of the system was conducted in a controlled laboratory environment. The primary objectives of the system were to (1) successfully obtain data from the sensing module, (2) obtain data from a second dummy sensing module without having to reprogram the controller node, and (3) package and send the retrieved data to the remote server.

Laboratory testing consisted of placing the water quality sensing probes into a glass container with water and activating the controller node through the Android phone's GUI. This would trigger the procedure indicated in the controller node's stored configuration file several times over the course of five minutes, at approximately 15 second intervals.

After activation, the controller node was able to successfully pull data from the water quality sensing module through a Bluetooth connection and display it through the phone's GUI. The remote server's frontend website was accessed to confirm that the data from the controller node was also being successfully recorded in the online database. This test was marked as a success. Fig. 5 shows part of the results obtained from the controller node's logs.

```
[08.51.41.795] pH: 5.79 DO2: 8.88 COND: 1646.0
[08.51.55.364] pH: 5.72 DO2: 8.88 COND: 1629.0
[08.52.08.884] pH: 5.63 DO2: 8.88 COND: 1637.0
[08.52.22.406] pH: 5.61 DO2: 8.88 COND: 1722.0
[08.52.35.980] pH: 5.57 DO2: 8.88 COND: 1620.0
[08.52.49.486] pH: 5.54 DO2: 8.88 COND: 1620.0
[08.53.03.033] pH: 5.5  DO2: 8.88 COND: 1615.0
[08.53.14.153] pH: 5.5  DO2: 8.88 COND: 1620.0
[08.53.27.681] pH: 5.46 DO2: 8.88 COND: 1620.0
[08.53.41.197] pH: 5.43 DO2: 8.88 COND: 1646.0
```

Fig. 5. Controller Node Data Log Sample

The researchers then conducted a second test wherein the dummy sensing module was activated simultaneously with the water quality sensing module. No new data appeared on the phone GUI over the span of three minutes, indicating failure.

A third test was also conducted with only the dummy sensing module active. As with the first test, the controller node was able to successfully pull data from the dummy sensing module, display it through the phone's GUI, and relay the data to the remote server. This test was also marked as a success.

7 Conclusion and Discussion

Based on the results of the first and third tests, it can be concluded that the framework is a viable design for WSN systems which have to continuously integrate new modules or swap out old ones without having to reprogram the core components of the system. As such, this framework can be used for semi-permanent remote installations that can be continuously upgraded over time with less expensive and more accurate sensing modules. Conversely, the core components such as the controller node and remote server can also be replaced with equivalent systems as long as they implement the communication protocol of the existing system's modules.

Other potential improvements can be to replace the current Bluetooth shields with low-energy Bluetooth so as to be more energy efficient, implement a module for power monitoring and switching to conserve energy, and redesign the communication protocol between the controller node and the system modules to be more and secure.

Acknowledgements. Funding for this research was provided by the Department of Science and Technology Grants-In-Aid Program and was supported in part by the Ateneo de Manila University Research Council, the Commission on Higher Education (CHED) and the Philippine Higher Education Research Network (PHERNET).

References

1. World Health Organization – Water Resource Quality – Rivers, http://www.who.int/water_sanitation_health/resourcesquality/wqachapter6.pdf
2. World Health Organization – WHO Guidelines for drinking-water quality, http://www.who.int/water_sanitation_health/publications/2011/9789241548151_eng.pdf
3. World Health Organization – Water Resource Quality – Field Work and Sampling, http://www.who.int/water_sanitation_health/resourcesquality/wqmchap5.pdf
4. Martinez, K., Hart, J.K.: Environmental Sensor Networks. Computer 37(8), 50–56 (2004)
5. Ruan, Y., Tang, Y.: A Water Quality Monitoring System Based on Wireless Sensor Network and Solar Power Supply. In: 2011 IEEE International Conference on Cyber Technology in Automation, Control and Intelligent Systems (CYBER), pp. 126–129. IEEE (2011)
6. Rao, A.S., Marshall, S., Gubbi, J., Palaniswami, M., Sinnott, R., Pettigrovet, V.: Design of Low-Cost Autonomous Water Quality Monitoring System. In: Advances in Computing, Communications and Informatics (ICACCI), pp. 14–19. IEEE (2013)
7. Peng, J., Hongbo, X.: Design of a Water Environment Monitoring System based on Wireless Sensor Networks. Sensors 9(8), 6411–6434 (2009)
8. Nasirudin, M., Za'bah, U.N., Sidek, O.: Fresh Water Real-Time Monitoring System based on Wireless Sensor Network and GSM. In: 2011 IEEE Conference on Open Systems (ICOS), pp. 354–357. IEEE (2011)
9. Arduino Mega 2560, http://arduino.cc.en/Main/ArduinoBoardMega2560
10. Atlas Scientific Sensors, https://www.atlas-scientific.com/kits.html

Fun Learn: An Interactive Mobile Platform for Kids Learning

Shafiq ur Rehman

FAST National University, Department of Computer Science,
Islamabad, Pakistan
shafiq.rehman@nu.edu.pk

Abstract. The project brings out the utmost desire for kids to learn basic mathematics in the form of simple games. They can enjoy on one hand while on the other hand their mathematics skills can greatly be improved by using Fun Learn kids' games. Usually kids go more for playing games that make them feel entertained and are easy to play. This game will not only entertain the kids but will also make them learn unintentionally that makes it less boring than normal learning. The game would comprise of three basic games related to Math's learning. Score for a user will increase for every correct answer. The project adds some interesting design techniques to the levels which makes it even more attractive for kids to increment their urge to play it and quench their thirst through it.

Keywords: fun learn, interactive, user friendly, learning.

1 Introduction

Pakistan is the developing country that requires help and support in the field of education, especially for children learning to fasten up its pace. The current education system is not up to standard and need to improve the education system, so that better education system be implemented. There is a immense gap in rural and urban education system. On the other hand technology becomes everywhere, therefore we can get the benefit to implement the latest technology. This project offers mobile learning platform and hence fill the gap for rural and urban education system. It demonstrates and apply the latest mobile technology in the area of online education system.

Fun Learn, a game having fun and learning at the same time [5]. It is easy and it is entertaining. The game would comprise of three basic games related to Math's learning. These games include counting, finding smallest number and finding largest number. Players would select the game they wants to play. A number of questions would be given to the player in each game. Levels would be incremented upon giving the right answer. Marks would be increased each time there is an increment in level.

Games can be very productive learning activities [4]. If someone wants to find a way to keep their children entertained and facilitate them learn at the same time [3,8], consider FUN LEARN an android based application. In today's computerized society, parents and teachers struggle to make engaging projects that will help kids learn, so

K.J. Kim and N. Wattanapongsakorn (eds.), *Mobile and Wireless Technology 2015*,
Lecture Notes in Electrical Engineering 310, DOI: 10.1007/978-3-662-47669-7_12

FUN LEARN will be a natural choice for educational games as it will have a very fun learning and friendly interface. The motivation behind this project is to apply design techniques learned so far more effectively and in depth. The project will help us to come in accordance with different tradeoffs in design interface for more universality [1,2].

1.1 Objectives

- Kids will get themselves entertained on hand held gadgets.
- This application will provide opportunities for building conceptual skills and developing positive attitudes towards mathematics, by reducing the fear of error and failure.
- Intelligence of kids will start improving at small age and will help them to step up more confidently at the elementary education level.

2 Proposed Architecture

Following Figure 1 shows the proposed architecture of Fun Learn.

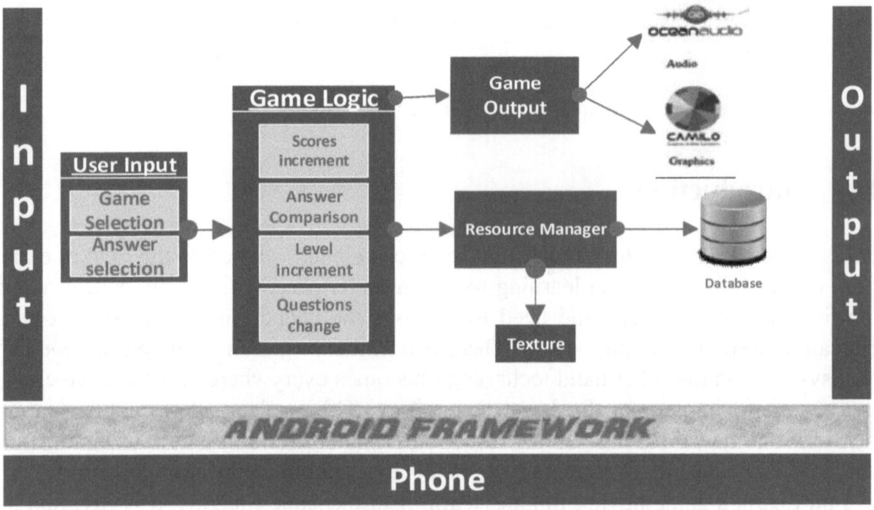

Fig. 1. Proposed Architecture

2.1 Description of Proposed Architecture

- **Input Module** will comprise of the selection of game out of three games including counting, finding smallest number and finding largest number. It will also consist of the answer selected by player to the given questions in selected game out of the three games.

- **Logic Module** is basically the business logic layer and is comprise of the game logic i.e.
  - The **increment in levels** when the player gives the right answer to the given question in the specific level of selected game.
  - **Increment in scores** upon giving the right answer to the given question when the level is incremented.
  - **Comparison** of the answer to the given question selected by the player with the correct answer from the database.
  - **Random change** in questions of each game in Fun Learn
- **Database** will have data of questions of the different games in Fun Learn and their correct answers.
- **Graphics** module will contribute in the visual and audio features of the game.
- **Resource Manager** manages the resources like texture and is connected to the database.

3 Prototypes

Children opens the application and a menu appears:

Fig. 2. Fun learn

- A cross icon at the top will close the game.
- This menu shows three types of games

1- Counting numbers
2- Find largest number
3- Find smallest number

- ➢ Select one of the options shown in figure 1

3.1 Counting Numbers

On selecting this option, questions about counting will be asked which will be solved by children.

Fig. 3. Count apple

- ➢ Figure 3 shows 6 apples and 3 options.
- ➢ Children have to count number of apples and choose the correct option
- ➢ Giving the right answer will lead to next level and the increment in scores.
- ➢ Similarly in this figure 4Children is asked to count the number of oranges
- ➢ Three options are given
- ➢ Giving the right answer will lead to the next level and increment in scores

Fig. 4. Count orange

3.2 Find a Smallest Number

On selecting this option, certain questions about finding the smallest number will be given to the Children.

Fig. 5. Find smallest number

- Children is asked to select the number which is smallest
- Giving the right answer will lead to the next level and scores will be incremented.

3.3 Find the Largest Number

On selecting this option, Children will be given multiple questions about finding the largest number.

Fig. 6. Find largest number

- Giving the right answer will lead the Children to the next level and scores will be incremented.

4 Conclusion

We have successfully implemented a mobile learning environment which is user friendly and easy to learn. Fun Learn is comprised of three types of games. Player selects one of the games according to his mood. It helps the little ones to enhance their conceptual as well as mathematics learning skills. Fun learn is easy and yet interesting and enjoyable to play with. The basic goal for making this game is to improve mathematics skills of children and to get themselves entertained at the same time.

References

1. Rehman, S., Coughlan, J.: Design Aesthetics of Mobile Interface. International Journal of Computer and Information Engineering (July 2012)
2. Holzinger, A., Nischelwitzer, A., Meisenberger, M.: Mobile phones as a challenge for m-learning: examples for mobile interactive learning objects (MI-LOs). In: Third IEEE International Conference on Pervasive Computing and Communications Workshops, PerCom 2005 Workshops, pp. 307–311. IEEE (March 2005)
3. Motiwalla, L.F.: Mobile learning: A framework and evaluation. Computers & Education 49(3), 581–596 (2007)
4. Facer, K., Joiner, R., Stanton, D., Reid, J., Hull, R., Kirk, D.: Savannah: mobile gaming and learning? Journal of Computer Assisted Learning 20(6), 399–409 (2004)
5. Kukulska-Hulme, A., Sharples, M., Milrad, M., Arnedillo-Sánchez, I., Vavoula, G.: Innovation in mobile learning: A European perspective. International Journal of Mobile and Blended Learning (IJMBL) 1(1), 13–35 (2009)
6. Halim, Z., Baig, A.R., Zafar, K.: Evolutionary Search in the Space of Rules for Creation of New Two-Player Board Games. International Journal on Artificial Intelligence Tools 23(02) (2014)
7. Halim, Z., Rauf Baig, A., Mujtaba, H.: Measuring entertainment and auto-matic generation of entertaining games. International Journal of Information Technology, Communications and Convergence 1(1), 92–107 (2010)
8. Rehman, S., Abid, M.: An Empirical Research on the User Acceptance of Mobile Payment System. In: International Conference on Information Society (i-Society 2014). IEEE (2014)

A Compressive Sensing Detection Approach Based on Spectral Kurtosis for Frequency Hopping Signal

Chenlin Hu[1], Jin Young Kim[1], Hyoung-Gook Kim[2], and Chang-Joo Kim[3]

[1] Chonnam National University, Gwangju, Rep. of Korea
beyondi@jnu.ac.kr
[2] Kwangwoon University, Seoul, Rep. of Korea
[3] ETRI, Daejeon, South Korea

Abstract. In order to determine the component of frequency hop, this paper taking the advantage of Compressive Sensing (CS) for sparse signal in frequency domain. Spectral kurosis is firstly calculated to identify peak locations which represent hopping frequency. Then the most optimal peaks are selected by applying Compressive Sensing Match Pursuit (CoSaMP) algorithm to overcome noise uncertainty. The hop time finally obtained from the residual between the given signal and the estimated one. The experiment results show that the system performs effectively even when SNR values are below -10dB, the presented approach can be effectively detect the hopping time. Compare to other conventional approaches, our proposed method is appropriate for lower and more widely varying SNR, reduces considerably the complexity.

1 Introduction

Communication security is an essential issue in various fields, especially in military and business transaction. Usually radio communication working under a fixed frequency, thence a series of problems appears, such as wiretapping, electronic warfare, or channel congestion. In order to avoid these problems, frequency hopping (FH) technique is utilized to assure communication's secrecy and immunity. In general, FH adapts the concept of changing a carrier frequency following a random sequence which only known by transmitter and receiver. Therefore, in practice as stated by L. Aydh and A. Polydoros [1], the hopping parameters, hopping time and frequency, should be estimated by receiver.

Over the last decade, the FH detection has attracted a great interest due to its effectiveness. There are many approaches were developed as described in [2], using autocorrelation techniques, based on image features, or based on atomic decomposition and information theoretic criterion. Meanwhile, Jia Yuan, Pengwu Tian and H. Yu [3] proposed a novel algorithm that taking advantage of CS for the sparse signal in frequency domain, where the peak represents hopping frequency. By calculating cosine of angles between incoherent measurements and column of Gaussian matrix, the maximum value indicates the position of hopping frequency. On the other hand, the blind detection [4] is used to determine the occurrence of frequency hop, accomplished by statistically analyzing the deviation between actual sampling values and the expectation.

K.J. Kim and N. Wattanapongsakorn (eds.), *Mobile and Wireless Technology 2015*,
Lecture Notes in Electrical Engineering 310, DOI: 10.1007/978-3-662-47669-7_13

Dealing with CS hopping detection, ones should consider number of frequency component N which is the essential factor of whole process. Most of previous works emphasize the importance of specifying N as well as their magnitude measured in the next stage. The details of these algorithms are revealed fruitfully in [5] such as Greedy algorithm including thresholding and Orthogonal Matching Pursuit (OMP). Conventionally, the l_0 or l_1-norm are adopted for the optimization process.

In this paper, we present a new approach of detecting signal's hopping time. We first select the most optimal peaks by calculating kurtosis and applying Compressive Sampling Match Pursuit (CoSaMP) algorithm [6] which is an enhancement of OMP. After that, the hop time can be obtained from the residual between the given frequency-domain signal and the estimated ones. Experimental results proved that the presented approach can be applied under noisy environments whose SNR is higher than -10dB. Compare to other traditional approaches mentioned in [7-8], our proposed method without measuring the energy of peaks, can be applied for lower and more widely varying SNR, reduces considerably the complexity.

2 Frequency Components Detection

2.1 Spectrum Peaks Detection

Given a discrete-time signal x, which has the 2048 length and an orthonormal basis Ψ, such as Fourier basis F [8]. Then signal x can be expressed as $x = \Psi s$, where s is the transformation of x. The compressed signal can be modeled as $y = \Phi x = \Phi F^{-1} s = \Theta s$, where Φ and Θ are random-sampling matrix and measurement matrix respectively. In our method, the CS rate is 0.15 which is proven in [9] to get a best result. After that, the peak candidates are generated by adopting the frequency-domain information of $\breve{s} = F\breve{y}$, where $\breve{y}$ is the extended version of y and $\breve{y}(n)$ is zero when n is not elements of CS time instances. An example of weighting coefficient $\breve{s}_l$ is illustrated in Fig.2.

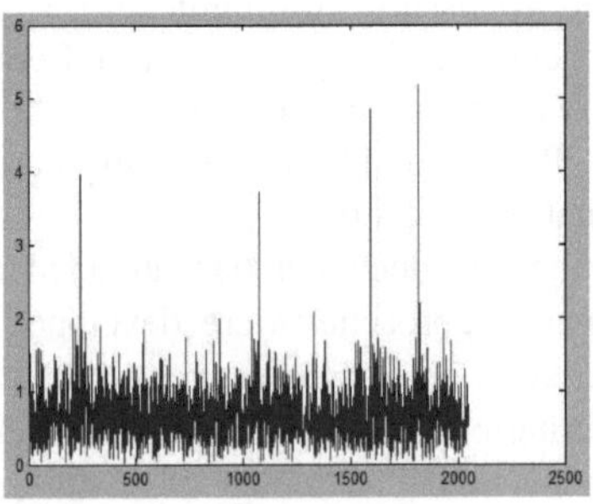

Fig. 1. The absolute value of $\breve{s}_l$ (CS rate 0.15, 5 frequencies components)

The distinct peaks in frequency-domain can be clearly observed in Fig.1 under preferable SNR. Then based on the signal energy detection, the number of peaks can be obtained. However, measuring the variance of noise (from time-domain and CS) and determining the threshold is not that of simple. Consequently, a method to measure noise and independent peaks is imposed.

Instead of exploring the peakedness based on Rayleigh distribution, we separately analyze the frequency in real and imaginary parts because Gaussian noise exists in both these two components independently. Fig.2 depicts the division process of $\breve{s}_l$ of signal from Fig.1.

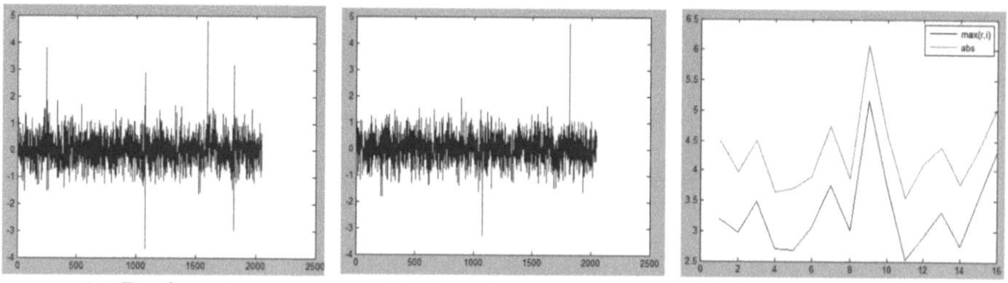

(a) Real part (b) Imaginary part (c) Comparison of l_1 or l_∞-norm

Fig. 2. Real and imaginary parts of $\breve{s}_l$

Apparently, the peaks in each component are more distinctive than that of the complex one. Therefore, we imposed the Kurtosis for each part separately and combine them either by l_1 or l_∞-norm Fig.2(c).

2.2 Kurtosis Calculation

Kurtosis is a measurement of the peakedness which describes the slope of candidates. The main frequency components are expressed under sharp peaks in real part, as well as the imaginary one. In addition, calculating kurtosis for the each 2048 candidates may lead the peaks to less contribution and increase the detection complexity. Therefore, we separate signal to 16 blocks, each has 128 samples.

2.3 Determine Value of K

With the determined threshold, in order to choose peaks as candidates which are higher and over threshold, we will follow the next steps.

a) Firstly, we divide $\breve{s}_l$ into $Q=16$ blocks and each block can be expressed as $\breve{s}_q, q=1,2,\cdots Q$
b) After that, an empty vector w for each block and a set P for storing peaks are initialized.
c) We calculate the kurtosis of each w vector. If it is smaller than the adaptive threshold, move to step e) and stop. Otherwise, we insert the value and position of the maximum w to set P. Then the value at w position is changed to zero.

d) We repeat step c) to collect other peaks in q^{th} block.

e) Output set P.

Finally K peaks are chosen as displayed in Fig.6 without noise.

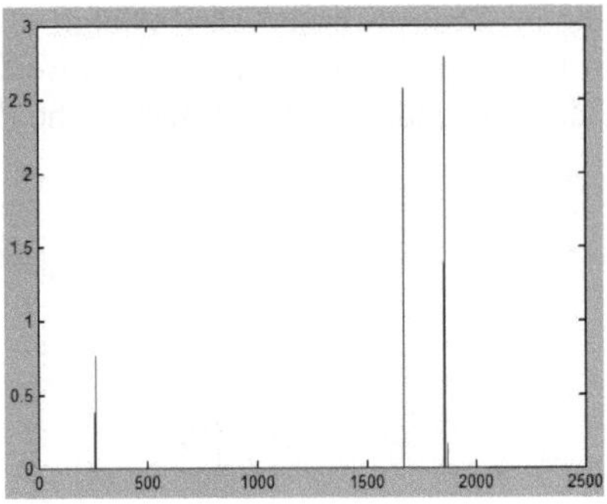

Fig. 3. K chosen peaks

Although kurtosis can determine spectral peaks, the detected candidates still encounters some false positive which is caused by the perturbation of high SNR noise. Therefore, we only utilize the number of peaks K in this stage. The exact positions of peaks are chosen by CoSaMP algorithm.

3 Frequency Hopping Time Detection

With the K detected peaks, the approximated signal can be reconstructed. Nevertheless, we only interested in detecting the hopping time. Therefore, instead of reconstruct the signal, we measured residual between approximation frequencies and ordinary one.

For $y(k)$ is the CS signal, $r(k)$ is the residual, then the index $rr(k)$ is utilized in the research.

$$rr(k) = \frac{|r(k)|}{|y(k)|} \tag{1}$$

In addition, we imposed the median filter described in [10] with rr to enable efficient crossing point detection. The impacts of this procedure are shown in Fig.4. The unfiltered residual, on the left side, is extremely noisy, which is impossible to detect the intersection. Otherwise, the one on the right included the median filter shows much enhancement. A conclusion can be deduced by observing the residual of specific frequency. Whether the hopping occurs in this frequency, the residual will be separated into two distinct segments surrounding hopping time. One part has large error, indicating the absence of this frequency, while the other tends to be much smaller.

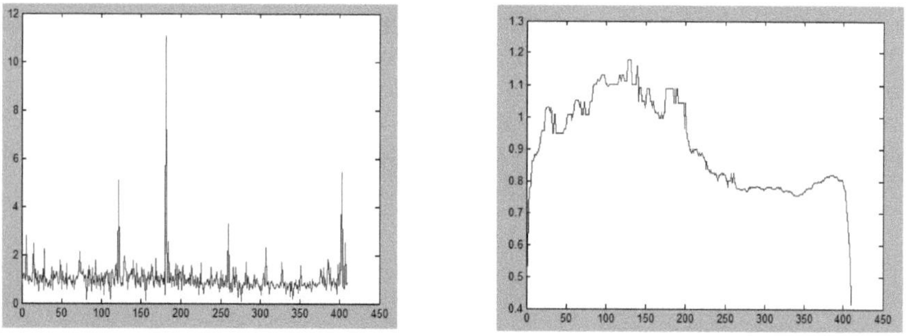

(a) Residual without median filter (b) Residual with median filter

Fig. 4. The median filter effects to residuals

In a certain case, $rr(k)$ is shown in Fig.5.

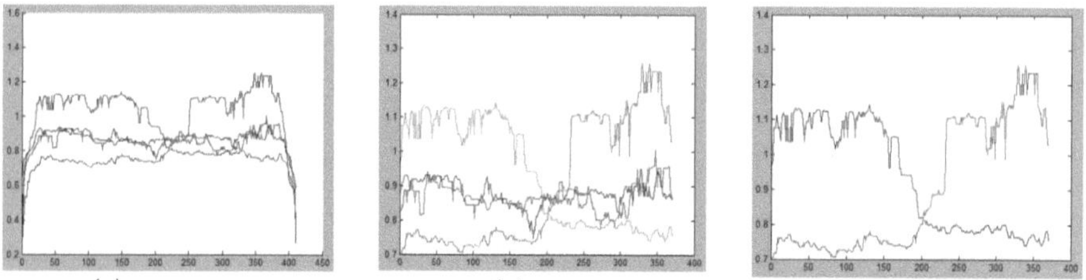

(a) $rr(k)$ of 4-sparse signal (b) $rr(k)$ without edge (c) Two hopping frequency components

Fig. 5. Residuals of approximation and original signal

Fig.5 illustrates the residual collected by adopting our proposed method. Obviously, there are 4 lines represents 4 distinct frequencies. Moreover, two of them intersect each other at the crossing point whose amplitude is around 1. In conclude, Fig.8 describes the case of compressing 4 frequencies signal, and the hopping time stays at the crossing point.

Given detected frequencies, a list of measuring CS candidates is generated by applying the similar CS scheme to their corresponding time domain signals. We argue that residual quantity, measured by the difference between the received CS signal and specific candidate, indicates the appearance of a target frequency. Through empirical observations, hopped frequencies usually contain a slope in residual line which represents the existence period. Specifically, the range of residual in which our interested frequency exists usually has smaller value than that of non-functional one. Otherwise, without hoping, the residual is expressed as a straight line showing a constant value in the whole frame

Supposing that our signal has 4-sparse including 2 constant frequencies and other 2 hopped ones, the two relative flat lines which are blue and green in Fig.5(b) represent the unchanged frequencies. Meanwhile, the cyan and red ones contain a slope and cross each other showing the occurrence of hopping event. Therefore, we select these two hopped frequency components which are blue and green in Fig.5(c) for analyzing.

After noticing this, we attempt to find the crossing point of the two lines. The two lines with larger difference, their crossing point is also the point has smallest difference of these two lines.

4 Experimental Results

To demonstrate the efficiency of our proposed approach, we conduct the experiments is to qualify our proposed approach for detecting hopping time under various noise models. The simulation is based on a 4-sparse FH signal with length is 2048, compressed with 0.15 CS rate and the hopping time is 1024.

We divide our qualification procedure into two steps due to the dependence of hopping time and the number of frequency. The purpose of first experiment is to analyze an adequate SNR which allows our method to successfully detect the hopping time. Then, the precision of hopping time estimation regards to those feasible SNR is further explored in the second step.

Table 1. Hopping time detection rate with different SNR

SNR	-20	-15	-10	-5	0	5
Detection rate	15.38%	30.77%	100%	100%	100%	100%

According to Table 1, our method can perfectly localize the hopping time derived from signal with SNR larger or equal -10dB. The poor performance, hopping time cannot be detected, is only appeared under the extremely noisy environments such as SNR smaller than -15. For further justification, we also explore the precision of the proposed method under the appropriate frequencies. The precision measurement is acquired using Equation (2) and the results are explained in Table 2.

$$\Pr ecision = 1 - \frac{|Detectedho\ ppingpo\ \mathrm{int} - 1024|}{1024} \times 100\,\% \tag{2}$$

Table 2. The precision of hopping time detection

SNR	-10	-5	0	5	10
Precision	99.76%	99.81%	99.85%	99.82%	99.84%

The results proved that our approach can precisely detect hopping time in environments with SNR is higher than -10dB. During the experiments, even SNR drops below -10dB, which has only 30.77% or 15.38% in detection rate, the estimated hopping times are always nearby 1024.

5 Conclusion

This study proposed a method to detect hopping time in frequency hopping signal using kurtosis and CoSaMP algorithm. Firstly, we compress the signal by Compressive Sensing method with CS rate is 0.15, and calculate kurtosis of each block which has the length of 128. Secondly, based on the results of kurtosis, we determine the number of frequencies K, and apply CoSaMP algorithm for selecting K optimal peaks from the given signal. Finally, the hopping time is obtained by analyzing residual between approximation frequencies and ordinary one. Experimental results explained that, the hopping time is 100% detected in the environment where SNR is higher than -10dB and the precision rate reach 99.8%. Compare to other traditional approaches, our proposed method without measuring the energy of peak, can be applied for lower and more widely varying SNR, reduce considerably the complexity.

Acknowledgments. "This work was supported by the ICT R&D program of MSIP/IITP [11-911-01-112, Technology Development for the My-F Convergence Service]."

References

1. Aydh, L., Polydoros, A.: Hop-Timing Estimation for FH Signals Using a Coarsely Channelized Receiver. IEEE Trans. Communications 44(4), 516–526 (1996)
2. Lopez-Risueno, G., Grajal, J., Yeste-Ojeda, O.A.: Signal detection and estimation using atomic decomposition and information-theoretic criteria. In: Proceedings of the IEEE International Conference on Acoustics, Speech, and Signal Processing (ICASSP 2004), vol. 2(17-21), pp. 1097–1100 (2004)
3. Yuan, J., Tian, P., Yu, H.: The Identification of Frequency Hopping Signal Using Compressive Sensing. In: Scientific Research Communications and Network, pp. 52–56 (2009)
4. Wu, J., Liu, N., Zhang, Y., Shen, C.: Blind Detection of Frequency Hopping Signal Based on Compressive Sensing, pp. 1691–1694 (April 2012)
5. Baraniuk, R.G.: Compressive Sensing. IEEE Signal Processing Magazine, 118–124 (July 2007)
6. Needell, D., Tropp, J.A.: CoSaMP: Iterative signal recovery from incomplete and inaccurate samples. Applied and Computational Harmonic Analysis 26(3), 301–321 (2009)
7. Lei, Y., Zhong, Z., Wu, Y.: A New Hop Duration Blind Estimation Algorithm for Frequency-Hopping Signals. Computational Intelligence and Industrial Application 1, 695–699 (2008)
8. Sirotiya, M., Banerjee, A.: Detection and estimation of frequency hopping signals using wavelet transform. In: Cognitive Wireless Systems (UKIWCWS), pp. 1–5 (December 2010)
9. Yu, N., Li, Y.: Deterministic construction of Fourier-based compressed sensing matrices using an almost difference set. EURASIP J. Adv. Signal Process. 2013(1), 1–14 (2013)
10. Hu, C., Kim, J.Y.: Spectral Domain Noise Modelling in Compressive Sensing-based Tonal Signal Detection. IEICE

Packet Loss Concealment for Improving Audio Streaming Service

Jun-Yong Lee[1], Hyoung-Gook Kim[1], and Jin Young Kim[2]

[1] Kwangwoon University, Seoul, Rep. of Korea
[2] Chonnam National University, Gwangju, Rep. of Korea
{jasonlee88,hkim}@kw.ac.kr,
beyondi@chonnam.ac.kr

Abstract. In this paper, we propose packet loss concealment for improving audio streaming service. Our approach is based on a concealment of lost packets and adaptive playout-buffer scheduling using active jitter estimation. The experimental results show that the proposed method enables users to deliver higher quality music.

1 Introduction

The use of the internet, Wi-Fi and mobile devices for transmission and reception of music, wideband speech, broadcast audio and podcasts is pervasive. Audio transmissions in real time over the internet are known as streaming media [1].

The promise of high quality of audio streaming is an important issue from a smart phone application viewpoint. However, since internet delivery doesn't guarantee quality of service, data packets are often delayed or discarded during network congestions, causing gaps in the streamed media. Previous works start from simple techniques, packet repetition [2], on to advanced techniques that use interpolation in the compressed domain for MPEG audio coders [3], [4].

In this paper, we propose a new receiver-based enhancing method of audio streaming quality that can be used to solve the problems described above. Our algorithms have two important improvements: 1) using accurate jitter estimation, our playout-buffer scheduling algorithm makes it possible to trade-off the buffering time with the rate of packet loss. 2) our packet loss concealment algorithm is based on alleviates the metallic artifacts that are often introduced during concealing consecutive packet loss.

2 Proposed Packet Loss Concealment for Improving Audio Streaming Services

The receiving portion of a smart phone for music quality over IP networks is composed of eight modules: packet information stripping, jitter buffer, network jitter estimation, decoder, signal frame buffer, playout scheduling, packet loss concealment and digital-to-analog converter.

K.J. Kim and N. Wattanapongsakorn (eds.), *Mobile and Wireless Technology 2015*,
Lecture Notes in Electrical Engineering 310, DOI: 10.1007/978-3-662-47669-7_14

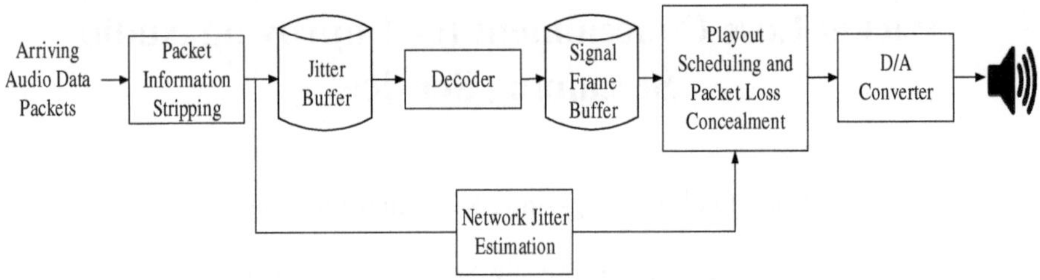

Fig. 1. Structure of the receiving portion of a smart phone

When an ith packet arrives at the receiver from the IP network, the receiver strips the packet information and places the packet adequately in the jitter buffer. The jitter buffer releases them for decoding at a regulated speed (i.e., every 10 ms), thereby reducing system delay. In response to the arriving packets in the jitter buffer, the network jitter estimation incorporates rapid detection of delay spikes to react to changes in network conditions.

Each signal frame decoded from the jitter buffer is used in packet loss concealment and playout scheduling which plays each packet at the scheduled time in order to prevent the degradation of the audio quality using the normal operating or compressing of each packet according to the network jitter estimated from the jitter. A digital-to-analog converter regularly converts the processed signal frame at a scheduled time (i.e., every 20 ms) into an analog signal. Finally, the user hears the analog audio signal through a speaker.

2.1 Packet Loss Concealment

Figure 2 presents an overall flow chart of packet loss concealment. The main idea of loss concealment is generating a signal that has a pitch period close to the original signal using the pitch period information of the past and future frames, thereby minimizing boundary discontinuities between the good packets and the reconstructed packets.

The frames most similar to the lost one are past or future signal frames. For this reason, the $(i\text{-}1)^{th}$ signal frame is searched in the history buffer. The signal frame has a length of 20 ms. The last N samples from the $(i\text{-}1)^{th}$ signal frame are selected as reference segment S, which is used to find similar signal segments in the past and future signal frames. During concealment of the missing packets, two different situations can arise: 1) the $(i+1)^{th}$ future packet or $(i+1)^{th}$ future signal frame is not available; or 2) the $(i+1)^{th}$ future packet or $(i+1)^{th}$ future signal frame is available.

If the $(i+1)^{th}$ future signal frame is not available (the $(i+1)^{th}$ future packet or $(i+1)^{th}$ signal frame is not found in the signal frame buffer and in the jitter buffer) for reconstruction of lost packets, the concealment algorithm works only on the past N samples (of the $(i\text{-}1)^{th}$ previous signal frame) stored in the history buffer. The area from the most similar segment, H, to the reference segment, S, is searched in the history buffer. The search is done using a normalized cross-correlation measure, and the range is limited to a pitch-compatible range. The ith lost signal frame is substituted by direct repetition of the detected signal segment H. If the $(i+1)^{th}$ future signal frame is available, (the $(i+1)^{th}$ future signal frame is found in the signal frame buffer or the $(i+1)^{th}$ future packet is present in the jitter buffer), the concealment algorithm uses the $(i+1)^{th}$ future packet information (or the $(i+1)^{th}$ future signal frame from the signal frame buffer) and past signal frame information from the history buffer for the lost packet reconstruction.

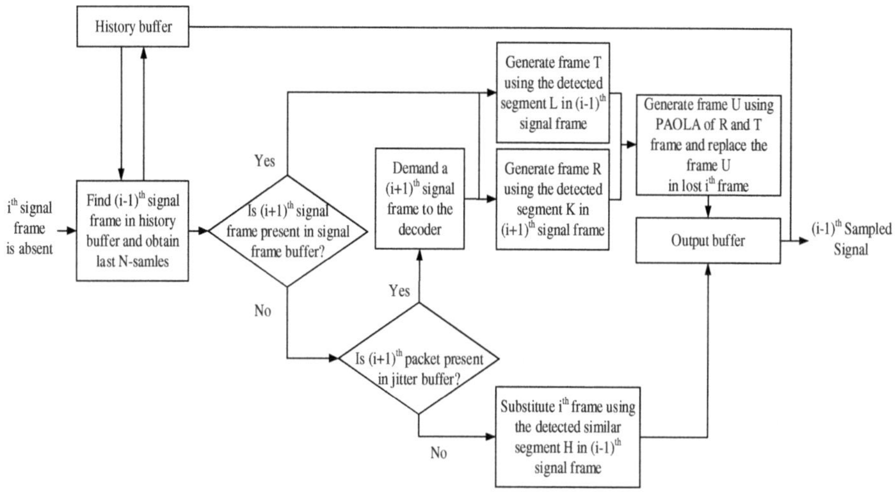

Fig. 2. Overall flow chart of the packet loss concealment

The refined estimated signal frame U for substituting the lost signal frame results in a smooth interpolation using peak alignment overlap-add (PAOLA) [5] between the signal frame R and the signal frame T. The efficient PAOLA algorithm uses a simple peak alignment technique to synchronize synthesis frames and takes waveform properties and the desired time-scale factor into account to determine the optimal parameters. The reconstructed signal frame U smoothes the discontinuity that affects the perceived quality of the reconstructed signal frame.

3 Experimental Results

For the experiment, we collected five network delay traces from the Internet links. For measuring music quality, we used a MOS experiment. In the MOS experiment, 24 listeners (12 male and 12 female students) were involved. We asked subjects to rate the listening quality of thirty musical songs (twenty-second long sequence for each song) using a 5-point scale: 1-bad, 2-poor, 3-fair, 4-good, and 5-excellent. Table I depicts the experimental results.

From Table 1, the highest MOS scores (> 4.0) were obtained in traces 1 and 2, which do not carry any network losses. In particular, as the jitter levels or packet loss rates increase, the MOS scores decrease.

Table 1. Experimental results of five network delay traces

Trace	Maximum Jitter (ms)	Network packet loss (%)	MOS
1	48	1.95	4.1
2	50	4.02	4.0
3	157	3.99	3.8
4	199	2.12	3.9
5	366	1.98	3.7

4 Conclusion

An audio streaming method based on packet loss concealment was proposed and evaluated. The packet loss concealment improves the audio quality over the IP networks and enables users to deliver higher quality music. The experimental results encourage the use of the proposed fully receiver-based enhancing algorithm in many practical mobile audio streaming applications. In future work, we will apply the method to spatial audio reproduction systems running on mobile phones.

Acknowledgments. This research was supported by the MSIP(Ministry of Science, ICT and Future Planning), Korea, under the ITRC(Information Technology Research Center) support program (IITP-2015-H8501-15-1016) supervised by the IITP(Institute for Information & communications Technology Promotion).

References

1. Pedram, S.K., Vaseghi, S., Langari, B.: Audio Packet Loss Concealment using Motion-Compensated Spectral Extrapolation. In: Signal Processing and Information Technology (ISSPIT), Athens (2013)
2. Mpeg-1: Coding of Moving Pictures and Associated Audio for Digital Storage Media at up to 1.5 mbit/s, Part 3: Audio. International Standard, IS 11172-3, ISO/IEC JTC1/SC29 WG11 (1992)
3. Lauber, P., Sperschneider, R.: Error Concealment for Compressed Digital Audio. In: AES 111th Convention, New York (2001)
4. Quackenbush, S., Driessen, P.: Error Mitigation in Mpeg-Audio Packet Communication Systems. In: AES 115th Convention, New York (2003)
5. Dorran, D.: Audio Time-scale Modification. Dublin Institute of Technology Doctoral Thesis (2005)

Secure IP Mobility Support in Software Defined Networks

Shimin Sun, Li Han, and Sunyoung Han*

Department of Computer Science & Engineering, Konkuk University
Seoul, 143-701, Korea
{shimin,hanli,syhan}@cclab.konkuk.ac.kr

Abstract. With the proliferation of mobile device and applications, a large number of mobility management solutions have been proposed to support mobility in IP networks. Mobile IP and its derivatives are basically network layer solutions for terminal mobility problems. However, there are several shortages in those solutions, such as large overhead, high latency, lack of scalability and weak security. Software Defined Networking (SDN) provides a flexible centralized approach to handle mobility in IP networks. In this paper, we propose a network based secure mobility management architecture. SDN controller manages the mobility with full authority and controls OpenFlow Switches to forward packets for mobile nodes in optimal path. The implementation and test results show a better delay and throughput performance during the handover process.

1 Introduction

With the growing number of mobile devices (e.g., smart phones, tablets, and laptop computers), the demand for mobility management is becoming a primary requirement in current mobile infrastructure. Today's cloud-enabled, everything-as-a-service (XaaS) [1] network environments expect to provide simple and secure mobile access to all the data and applications for users, regardless of their physical location. Service providers want to make it easy to locate users and provision their services ubiquitously. Currently, IP mobility solutions (Mobile IP (MIP) [2] with its derivatives, variations, and auxiliary [3-6]) are insufficient in terms of scalability and resource utilization to large scale mobile networks. It is urgent to excogitate more flexible and easier methods to provide world-wide mobility support.

Software Defined Networking (SDN) [7] is an emerging network architecture approach. The separation of control plane with data plane is its essential feature and the evolution comparing with traditional fixed IP networks. In SDN, the underlying network infrastructure is abstracted from applications while network structures, functions and status are logically centralized. OpenFlow [8] is the potential protocol for the communication between controllers and OpenFlow Switches (OFSs). OFSs are simply responsible for data forwarding based on their flow table which is configured by controller. SDN controllers expose and abstract various network functions and

* Corresponding author.

K.J. Kim and N. Wattanapongsakorn (eds.), *Mobile and Wireless Technology 2015*,
Lecture Notes in Electrical Engineering 310, DOI: 10.1007/978-3-662-47669-7_15

operations to OFSs via its programmatic interfaces. Some benefits can be taken by introducing SDN concept into mobile networks, including resource utilization optimization, heterogeneous network interconnection, load balance and QoS management. As we present in this paper, IP mobility support could also be realized with new network functions in software defined approach. A flow-based paradigm ideally suits for mobility management in high scalable mobile networks.

The purpose of mobility support is to maintain session continuity from application's perspective. Normally, when a mobile node (MN) changes its access point from its home network, it gets a new IP address from the new attached network. On-going sessions are interrupted due to the change of IP address. To afresh the connection between MN and its corresponding nodes (CNs), new sessions need to be initiated with the new IP address of MN. To keep the connectivity of on-going session, some mobile IP protocols have been proposed. Mobile IPv4 (MIPv4) [3] or Mobile IPv6 (MIPv6) [4] allocate an IP address, named Care-of address (CoA), to MN as the locator in the new network. The original IP address of MN, named Home Address (HoA), performs as identifier which is continually used for the on-going sessions. An indirection agent, called Home Agent (HA), stores binding cache of each MN in the form of a mapping between MN's HoA and its current CoA. In addition, the tunneling or IPv6 extended headers are used to keep sessions between HA and Foreign Agent (FA). All packets from Corresponding Nodes (CN) to MN have to take a detour to HA first and then HA redirects those packets to the CoA of MN, which causes triangular routing [9]. Triangular routing leads to high overhead and heavy load on HA with non-optimal path. Some countermeasures [10-12] have been proposed to alleviate the impact of triangular routing.

Security in mobile networks is one of the most critical challenges due to the wireless environment and unpredictable connection requirements from authorized or unauthorized MNs. The authentication outlined in MIP only serves to authenticate MN with HA. The lack of authentication of MN with FA and FA with HA is the vulnerable points. In SDN, the nature security consideration provides the advance authentication between controller and OFSs. Authentication is necessary only between controller and MN. In this paper, we use a simple authentication algorithm to authenticate MN with controller. After authentication, controller executes a series of actions to achieve mobility support, which is elaborated in Section 3.

In this paper, we argue that SDN helps to offer secure mobility support in future SDN-enabled Internet architecture. It has several advantages to design OpenFlow based IP mobility support. Firstly, the centralized management architecture is beneficial to reduce protocol complexity. Without the support of any mobile IP protocols, node mobility can also be achieved. Secondly, the triangular routing can be eliminated by optimizing the routing path among OFSs. Thirdly, no HA or FA is required in this architecture. Controller and OFSs realize all the functions for mobility support.

The paper is organized with following structure: Section 2 is an overview of the literature on mobility. The proposed architecture for IP mobility and authentication mechanism are elaborated is Section 3. Prospective of performance is evaluated by comparing with MIPv6 in Section 4. Finally, conclusion and some indications of future work are presented in Section 5.

2 Related Work

IP Mobility management is an important research topic for current and future mobile network. A number of mobility management technique and architectures have been proposed and developed in order to provide the optimal mobility support. Most IP mobility solutions are developed based on MIP and its derivations. Some architectures separate identifier and locator to support terminal mobility. Recently, studies on IP mobility in SDN have been activated to construct new architectures or to improve the performance of existing solutions.

2.1 Mobile IP and Its Derivatives

MIP has been proposed by IETF to provide global mobility. For particular purposes, many approaches are proposed extending or optimizing the architecture of MIP. Hierarchical Mobile IPv6 (HMIPv6) [5] allocates Mobility Anchor Points (MAP) to localize MNs. MAP can significantly decrease the latency caused by binding update after handover, especially when HA is far away from home network. Proxy Mobile IPv6 (PMIPv6) [6] is a similar protocol architecture, and it is completely transparent to MNs from mobility and deploys Mobile Access Gateways (MAG) to perform mobility management on behalf of MNs. However, both HMIPv6 and PMIPv6 need to establish tunneling between HA and MN or between HA and MAG, so that triangular routing cannot be avoided.

Mobile IP and its derivatives don't scale well with a large number of MNs because of the significant overhead. Some research [11] [12] propose to allocate distributed HAs over the network. HAs use specific routing protocol to form an overlay network that allows them to exchange mobile nodes' data traffic. It obviously requires complex protocols to realize related functions, such as HA discovery, HA switching, and binding update synchronization. In order to reduce the overhead for inter-domain mobility, the distributed mobility management (DMM) [13] has been proposed, which allocates FAs in distributed manner and exchange location information of MNs with each other.

2.2 Other Network Layer Mobility Solutions

Some interesting and promising approaches behind the mobility management are to introduce new abstract identifiers to MNs. The identifier, could be host identity tag (HIT) described in Host Identity Protocol Version 2 (HIPv2) [14], is used for connection identification instead of IP address. Therefore, the change of IP address does not affect the continuity of its on-going sessions. Based on HIP, Locator/ID Separation Protocol (LISP) [15] was developed to improve the flexibility and scalability of IP routing. LISP uses Endpoint Identifier (EID) and localization information Route Locator (RLOC) replacing IP address but in the form of IP address. To handling mobility based on LISP, LISP-Mobility [16] inserts an abstraction layer between the network and transport layer. ID-based mobility solutions need to modify IP network models, which is unpractical to current Internet architecture. In future Internet research field, MobilityFirst [17] has been proposed, where every network endpoint as well as data content has a global unique ID to solve the issue of effective support of mobility.

2.3 SDN-Related Mobility Solutions

Along with the emerging of SDN, some researchers initiated the studies of mobility support in SDN. Literature [18] implements MIP within SDN networks to eliminate tunneling and perform route optimization. A pure SDN based IP mobility solution has been designed in [19] and strengthened in [20]. In this solution, OFSs in terminal network performs as FA to issue CoA for MNs. Controller has full authority to manage the mobility by store binding updates and achieves mobility by modifying flow tables of some OFS on the path from CN to MN. The shortages of this solution are that OFSs need to import the generation function of CoA and new binding update messages need to be designed to OpenFlow protocol. Additionally, security is not considered yet in those approaches. There are still many interesting issues need to be solved to make mobility support in SDN, such as management architrecture, mapping between location and address, and security. In this paper, we attempt to achieve the mobility support by designing mobility management functions on controller without modifying OpenFlow protocol or introducing new functions OFSs.

3 Secure SDN Based Mobility Support

Agent or anchor based mobility protocols usually lead to non-optimal routing, traffic overhead, additional network agents, and single point of failure. Central controlled SDN provides the possibility to provide the mobility management abnegating the MIP architecture. We propose that controller is the above all network manager which not only provides the ability to control OFSs and but also can manage MNs by introducing new functions. Controller takes full charge of mobility of MNs, which manages packet forwarding for MNs by updating the flow tables of corresponding OFSs. MNs can be totally transparent to the change of network attachment.

3.1 Architecture Description

We put the mobility management function on top of the controller. MNs which have roaming demands should register to controller preliminarily. Public Key Infrastructure (PKI) is used to provide MN's identity authentication and session security. After handover of a MN, controller maintains a session table of this MN's on-going sessions. In this table, an entry contains: HoA of MNs, port number of MN for a session, IP address of direct connected OFS and an assigned port number of OFS. The entry is deleted when the session expired. MNs are transparent to the handover process, and unnecessary to require a new CoA. Controller manages OFSs to achieve the whole communication.

After Handover, instead of generating a new CoA to deliver messages to the Internet, we prefer to use the IP address of direct connected OFS to route messages from MN. The problem is that if there is more than one MN, it is difficult to use one IP address of OFS to route messages for multiple MNs. We imitate the idea of Network Address Translation (NAT). Since current OFSs don't support NAT, we propose to achieve similar function by controller. As described in OpenFlow protocol, if a message reaches an OFS with no matching flow entry in its flow table, this message is sent to controller by *packet_in* message. The meaningful result is that all the first messages of each session from MN to the new OFS are encapsulated in *packet_in* and sent to controller. Then controller can obtain the information of all sessions initiated from MN. These sessions (represented by HoA of MN plus port number) can be mapped to the IP address of OFS with an unused port of this OFS. Since OFSs are network devices which are only running network services, the occupied port and the well-known port should be aware by controller.

3.2 Work Principle of the Architecture

The detail of this architecture is described as follows. How MN reaches CN is illustrated in Fig. 1 and how CN reaches MN is illustrated in Fig. 2. We assume that the HoA of MN, IP of visited OFS (S3), and the IP of CN are IP_{MN}, IP_{S3} and IP_{CN} respectively. S1 is the OFS located at the network of CN, while S2 is the OFS in home network of MN and S3 is the visited OFS. The port number of a session of MN is $PORT_{MN}$ and the mapping port of OFS is $PORT_S$. Then an entry in session table of controller is (IP_{MN}, $PORT_{MN}$, IP_{S3}, $PORT_{S3}$).

First, we describe the initial message that is sent from MN to CN through the visited OFS (S3). When the message reaches S3, S3 has no matched flow entry. So it encapsulates the packet and sends to controller via *packet_in* message. If it is the first message from MN, controller need to authenticate the MN primarily using the algorithm discussed in Section 3.3. If the authentication is success, an entry (IP_{MN}, $PORT_{MN}$, IP_{S3}, $PORT_{S3}$) is added into the session table in controller. After that, controller adds a flow table entry to S3 by *flow_mod* message, which indicates "for all message with source address IP_{MN} and source port $PORT_{MN}$, rewrite their source address to IP_{S3} and source port to $PORT_{S3}$". Meanwhile, one *flow_mod* message is sent to S2 to deletes all the flow table entries that destination is MN: "for all message with destination address IP_{MN}, delete the entry". Another *flow_mod* message is sent to S1, which indicates "for all message with source address IP_{S3} and source port $PORT_{S3}$, rewrite their source address to IP_{MN} and source port to $PORT_{MN}$". The flow from MN to CN will go through three redirections: MN → S3 → S1 → CN. In addition, the flow table entry for the reverse path from CN to MN should be added to S1 and S3. For S1, the entry is "for all message with destination address IP_{MN} and port $PORT_{MN}$, rewrite their destination address to IP_{S3} and port to $PORT_{S3}$". For S3, the entry is "for all message with destination address IP_{S3} and port $PORT_{S3}$, rewrite their destination address to IP_{MN} and port to $PORT_{MN}$". The whole process is illustrated in Fig. 1.

Next we describe the communication initiated by CN after handover, as shown in Fig. 2. After handover, MN may keep silent and may not initiate any messages destination to CNs. However, the attachment of MN to S3 can be perceived by potential messages, such as ARP, ICMP, Neighbor Discovery and Multicast Listener Discovery. Once the controller perceives the new attachment of MN, after authentication, it sends a *flow_mod* message to S2 which is the OFS at MN's home network. This *flow_mod* message deletes all the flow table entries that destination is MN: "for all messages with destination address IP_{MN}, delete the entry". Until now, the message from CN to MN is still routed to MN's home network. When receives the message destination to MN, S2 redirects the message to controller in *packet_in* messages. Controller creates an entry in session table with (IP_{MN}, $PORT_{MN}$, IP_{S3}, $PORT_{S3}$). A *flow_mod* message is sent to S2 to do the action of redirecting this and following messages from same session to S3.

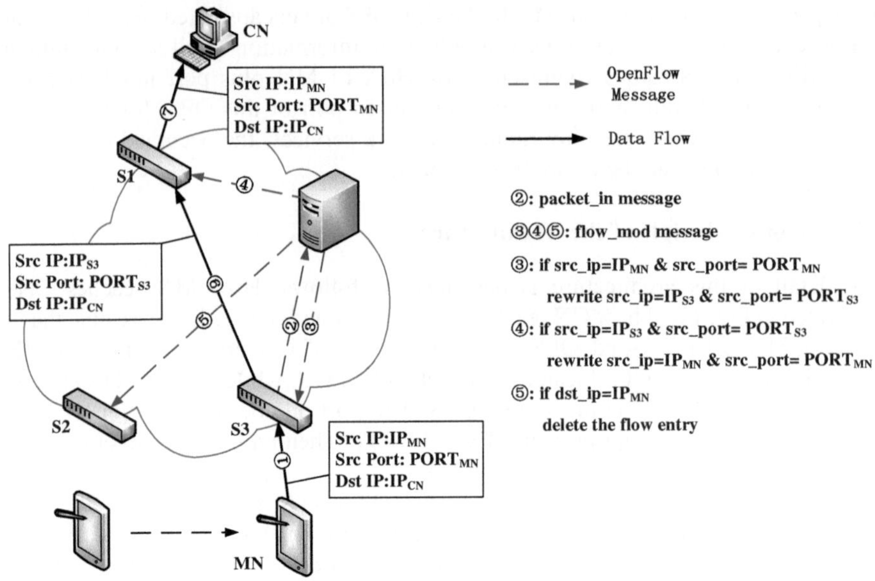

Fig. 1. First message of a session initiated by MN after handover

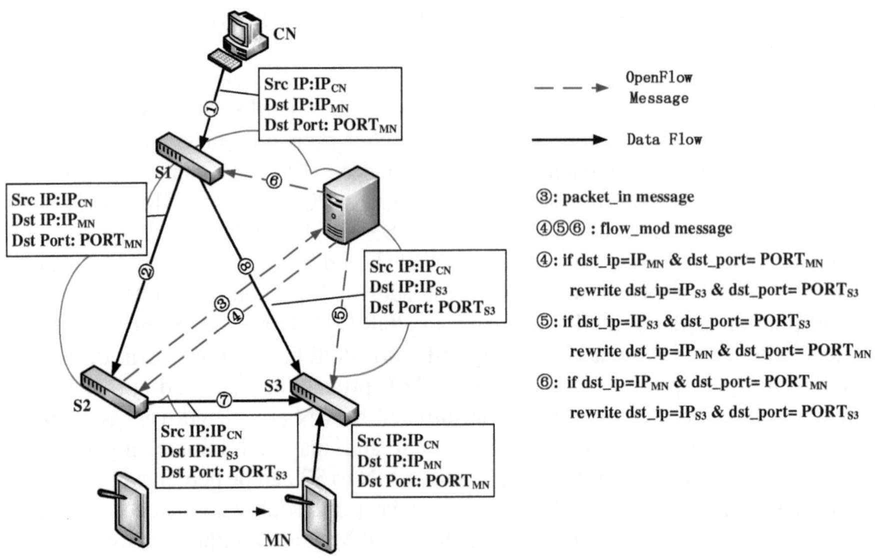

Fig. 2. First message of a session initiated by CN after handover

Another two *flow_mod* messages are sent to S1. One is to delete all MN related entries in flow table so that other messages with destination to MN will be sent to controller first. The other *flow_mod* message is used to add an entry to in flow table: "for all message with destination address IP_{MN} and destination port $PORT_{MN}$, rewrite their destination IP to IP_{S3} and destination port to $PORT_{S3}$". In addition, a *flow_mod* message is sent to S3, which indicates "for all message with destination address IP_{S3} and destination port $PORT_{S3}$, rewrite their destination address to IP_{MN} and destination port to $PORT_{MN}$".

3.3 PKI Based Authentication Algorithm for Mobile Nodes

When a new MN connects to a network, the first thing is to verify the identity of this MN. Strong authentication mechanisms ensure MN belongs to the authorized entities. PKI is a good way to provide non-repudiation through digital signatures. Digital "identities" issued by trusted third parties Certificate Authority (CA) with a pair of public/private keys. The ID of each node could be IPv4 or IPv6 address. To achieve the security modular, controller and MNs should be registered at Trusted Authority (TA) and keep the updated certification by themself. We assume that controller should have the list of authorized MNs previously or can require the property of unknown hosts from an authority.

The process of authentication for MNs is described as follows. When controller receives a message from an unknown IP, it checks whether this IP belongs to an authorized MN or not. If this IP belongs to an authorized MN, this controller requests the public key of this IP from CA. Then, controller generates a large random number RN (e.g. 128 bits) using BBS pseudo-random number generator [21]. This random number with controller's ID (ID_{CON}) encrypted by MN's public key K_{PU_MN} is sent by controller to MN through the insecure communication channel. MN decrypts the message using its private key and gets the random number RN, controller's ID_{CON}. Subsequently, MN encrypts this RN with its ID (ID_{MN}) using its private key K_{PR_MN} and sends the message to controller. Controller gets the random number by decrypting the message using MN's public key K_{PU_MN}. If the received RN is same to the generated RN, the authentication of MN is achieved and mark the MN as authorized MN. If the authentication is failure, controller may send a *flow_mod* message to the OFS which directly connects to MN to discard all the messages from this MN. Through the similar algorithm, MN can also authenticate the identity of controller.

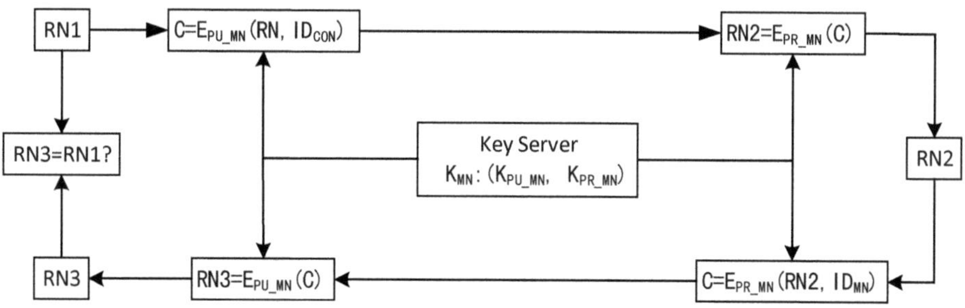

Fig. 3. Authentication process for mobile nodes

4 Implementation and Evaluation

Mininet [22] is the common used and convenient simulation tool for the implementation of SDN related researches. However, due to the limited OS capacity in mininet, it is difficult to deploy MIPv6 protocols. Therefore, we prefer to build the network topology in virtual environment using Virtualbox. The network topology is realized with 8 Linux virtual machines, including 1 CN, 1 MN, 5 OFSs (S1, S2, S3, S4 and S5) and 1 controller. The network topology is depicted in Fig. 4 (a), where S2 is the OFS at the home network of MN and MN moves from S2 to S4 passing by S3. To evaluate the performance of MIPv6, we installed MIPL [23] on MN, CN, HA, and FAs as shown in Fig. 4(b). We ran *iperf* on both MN and CN to monitor the variation of bandwidth during the evaluation. *Ping* command is used to get the variation of Round Trip Time (RTT) between MN and CN. The result is shown in Fig. 5.

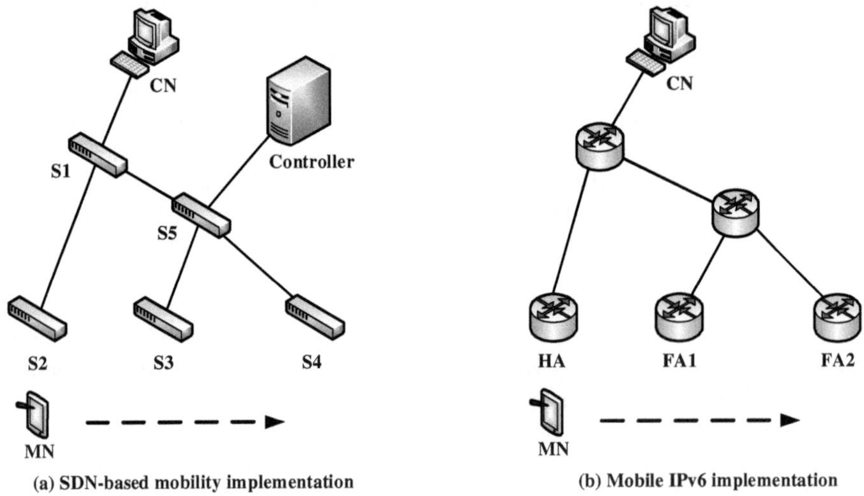

Fig. 4. Network topologies of the implementation

Fig. 5 (a) shows RTT of SDN based mobility is lower than MIPv6 in the same experiment scenario, where RTT values of both protocols temporarily raises to a higher value during handover process. RTT of MIPv6 stays at a higher value after handover. It is because when MN leaves home network, all packets to MN are relayed by HA results in triangle routing. The proposal avoids triangle routing by sending *flow_mod* messages to S1, S2 and S4 immediately after handover to build the new optimal path between MN and CN. Fig. 5(b) shows the variation of bandwidth between MN and CN tested using *iperf*. It can be seen that the bandwidth of the connection in MIPv6 decreases sharper than SDN based solution during handover process.

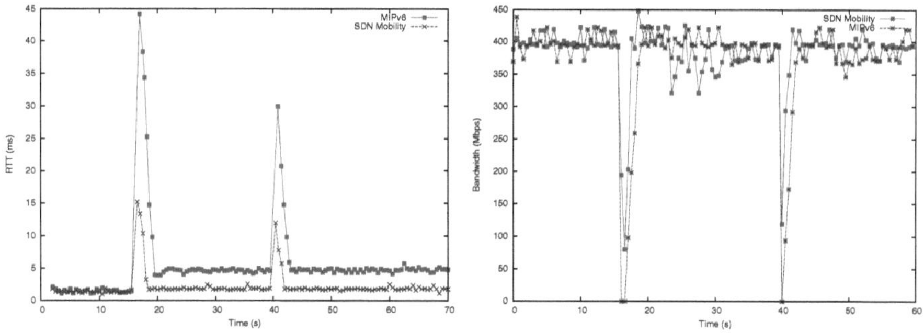

(a) Delay variation (b) Bandwidth variation

Fig. 5. Test results of delay and bandwidth variation during the period of MN moving from S2 to S4

5 Conclusion

Existing mobility management solutions are impractical with low scalability due to the requirement of auxiliary protocols or network agents. SDN is a new concept that has the opportunity to revolutionize the mobility support. Although similar concepts have appeared in the past with questionable success, the increasing requirement for network flexibility is expected to make SDN a promising mobility approach in mobile networks. In this paper, we proposed a novel mobility management framework based on SDN concepts. This framework provides more flexible solutions to manage networks than MIP based management protocols. To ensure scalability and feasibility of this solution for large mobile networks, distributed controller allocation, load balance and path optimization will be studied in the future.

Acknowledgments. This work was supported by ICT R&D program of MSIP/IITP, Republic of Korea [No. B0126151078002003, Creation of PEP based on automatic protocol behavior analysis and Resource management for hyper connected for IoT Services], China Scholarship Council [No. 201208260019 and No. 201308260028].

References

1. Banerjee, P., et al.: Everything as a Service: Powering the New Information Economy. Computer 44(3), 36–43 (2011)
2. Perkins, C.: IP Mobility Support. IETF RFC 2002 (1996)
3. Perkins, C.: IP Mobility Support for IPv4, Revised. IETF RFC 5944 (2010)
4. Johnson, D., Perkins, C., Arkko, J.: Mobility Support in IPv6. IETF RFC 6275 (2011)
5. Soliman, H., Castelluccia, C., ElMalki, K., Bellier, L.: Hierarchical Mobile IPv6 (HMIPv6) Mobility Management. IETF RFC 5380 (2008)
6. Gundavelli, S., Leung, K., Devarapalli, V., Chowdhury, K., Patil, B.: Proxy Mobile IPv6. IETF RFC 5213 (2008)
7. ONF White Paper: Software-Defined Networking: The New Norm for Networks. Open Networking Foundation (2012)

8. McKeown, N., et al.: OpenFlow: enabling innovation in campus networks. ACM SIGCOMM Computer Communication Review 38, 69–74 (2008)
9. Zhu, Z., Wakikawa, R., Zhang, L.: A Survey of Mobility Support in the Internet. IETF RFC 6301 (2011)
10. Sajal, S., Asish, K.M., Surajjan, G.: Fast Handoff Mobile IP Protocol Using A Novel Route Optimization Technique. Int. J. Recent Trends in Engineering 2(1) (2009)
11. Mao, Y., Knutsson, B., Lu, H., Smith, J.: DHARMA: Distributed Home Agent for Robust Mobile Access. In: Proc. of the IEEE Infocom (2005)
12. Cuevas, R., Guerrero, C., Cuevas, A., Caldern, M., Bernardos, C.J.: P2P Based Architecture for Global Home Agent Dynamic Discovery in IP Mobility. In: 65th IEEE Vehicular Technology Conference (2007)
13. Chan, H., Liu, D., Seite, P., Yokota, H., Korhonen, J.: Requirements for Distributed Mobility Management. IETF RFC 7333 (2014)
14. Moskowitz, R., Heer, T., Jokela, P., Henderson, T.: Host Identity Protocol Version 2. IETF RFC 7401 (2015)
15. Farinacci, D., Fuller, V., Meyer, D., Lewis, D.: The Locator/ID Separation Protocol (LISP). IETF RFC 6830 (2013)
16. Hertoghs, Y.H., Binderberger, M.B.: End Host Mobility Use Cases for LISP. IETF Internet-Draft, draft-hertoghs-lisp-mobility-use-cases-01 (2014)
17. Arun, V., James, F.K., Dipankar, R., Kiran, N.: MobilityFirst: A Mobility-Centric and Trustworthy Internet Architecture. ACM SIGCOMM Computer Communication Review 44 (2014)
18. Pupatwibul, P., Banjar, A., Sabbagh, A.A.L., Braun, R.: Developing an Application Based on OpenFlow to Enhance Mobile IP Networks. In: Local Computer Networks (LCN) Workshop on Wireless Local Networks (2013)
19. You, W., Jun, B.: A solution for IP mobility support in software defined networks. In: 23rd Int. Conf. Computer Communication and Networks, ICCCN (2014)
20. You, W., Jun, B., Keyao, Z.: Design and Implementation of a Software Defined Mobility Architecture for IP Networks. Mobile Networks and Applications 20(1), 40–52 (2015)
21. Sidorenko, A., Schoenmakers, B.: Concrete Security of the Blum-Blum-Shub Pseudorandom Generator. In: Smart, N.P. (ed.) Cryptography and Coding 2005. LNCS, vol. 3796, pp. 355–375. Springer, Heidelberg (2005)
22. Karamjeet, K., Japinder, S., Navtej, S.G.: Mininet as Software Defined Networking Testing Platform. In: Int. Conf. Communication, Computing & Systems (2014)
23. MIPL, Mobile IPv6 for Linux, `http://tldp.org/HOWTO/Mobile-IPv6-HOWTO/mipv6.html`

A Comparative Performance Analysis of MANET Routing Protocols under Security Attacks

Muhammad Saleem Khan, Qasim Khan Jadoon, and Majid I. Khan

COMSATS Institute of Information Technology,
Park Road, Islamabad, Pakistan
skhan.ciit@gmail.com, qasim_2009@live.com,
majid_iqbal@comsats.edu.pk

Abstract. This paper comprehensively analyzes the impact of security attacks on the performance of both proactive (OLSR, DSDV) and reactive (AODV, DSR) MANETs routing protocols. The analysis is provided in terms of performance metrics, such as average end-to-end delay, packet delivery ratio, and normalized routing load. Simulations are setup in NS-2 and performance of these routing protocols is discussed under gray-hole and black-hole attacks. Our finding shows that these protocols are designed without considering security aspects, as result, performance of these protocols degrades significantly under attacks.

1 Introduction

Mobile Ad-hoc Networks (MANETs) [1-2] is a group of self-organizing mobile nodes operating without any central authority. Node mobility often leads to dynamic network topology that makes routing in MANETs an important research area. Various routing protocols have been proposed in this regard that can be categorized into two primary categories i.e. Proactive and Reactive [3, 4, 5, 6]. Proactive routing protocols establish routing paths in advance for all source and destination pair. These routing paths are periodically advertised within the network to maintain up-to-date routes. Protocols in the aforementioned category are Destination Sequenced Distance Vector (DSDV) and Optimized Link State Routing Protocol (OLSR). Reactive routing protocols establishes on demand routing paths from source to destination. Dynamic Source Routing (DSR) and Ad Hoc On-Demand Distance Vector (AODV) are the examples of routing protocols in this category.

In *OLSR* [3], a routing table is maintained by each node that contains route information to each and every node in the network. In *DSDV* [4], nodes advertise their routing information including the address, sequence number and number of hops. In *AODV* [5] (RFC: 3561), when data transmission is required, source node broadcast route request (RREQ) packet to its neighbours. If the destination is found, unicast Route reply (RREP) is sent to the source. Whenever the failure occurs, route error (RERR) packet is sent towards the source node. *DSR* [6] also belongs to the reactive category of routing protocols. The route discovery mechanism of DSR is similar to

K.J. Kim and N. Wattanapongsakorn (eds.), *Mobile and Wireless Technology 2015*,
Lecture Notes in Electrical Engineering 310, DOI: 10.1007/978-3-662-47669-7_16

AODV, however, DSR uses source routing scheme where transmitted packets contain the complete route towards the destination. In DSR, each node caches a new route that it learns. Route caching speed up the route discovery mechanism, but the stale caches also affect the performance.

MANETs are inherently insecure due to the dynamic topologies, wireless links having limited capacity, interference problems, lack of both centralized control and established infrastructure. Moreover, in the design of MANETs routing protocols security is often not given due consideration. This paper provides a comprehensive evaluation of the four well-known MANET routing protocols (OLSR, DSDV, AODV, and DSR) and the impact of security attacks on their performance.

The paper is organized as follows, Section 2 discusses security attacks on MANETs routing protocols. Section 3 provides simulation setup and performance metrics. Simulation results and analysis are discussed in Section 4 and Section 5. Section 6 gives related work and Section 7 concludes the paper.

2 Security Attacks in MANETs

Security attacks in MANETs can be categorized into active attacks and passive attacks. In active attacks, malicious node make itself a legitimate node within the network and advertise fake routing information, such as shorter distance towards the destination or higher sequence number, to be the part of active route. In passive attacks, the attacker do not physically become part of the network but only listens to the on-going communication in the network. The goal is to get the specific information from the data rather than data modification or packet dropping. We have analyzed the impact of two active attacks Black-hole and Gray-hole on the performance of MANET routing protocols in this paper.

Black-hole Attack: In this type of attack, malicious node attracts traffic from all over the network by advertising false shortest route towards the destination and highest sequence number and then, drop all the packets forwarded through it [8, 9].

Gray-hole Attack: In this attack, the goal of the malicious node is to partially drop packets that are either destined to a particular destination or of selected traffic type [10]. Thus, the malicious node act normal while dealing with the general traffic type but as soon as it receives packet of specified type or destined to target destination, it drops those packets.

3 Simulation Setup and Performance Metrics

In this study, quantitative evaluation of routing protocols has been carried out using NS-2 simulator. We have analyzed the performance of four well-known routing protocols (a) OLSR, (b) DSDV, (c) AODV, and (d) DSR for following performance metrics under simulation setup shown in Table 1.

Packet Delivery Ratio (PDR): Ratio of sent data packets from source to data packets received at the destination.

Average End-to-End Delay: It is the average time taken by a data packet sent from source to destination.

Normalized Routing Load (NRL): It is the number of routing packets transmitted per data packet.

Table 1. Simulation Parameters

Parameters	Values	Parameters	Values
Pause Time	20s, 40s, 60s, 80s, 100s	Packet Size	512bytes
Data Rate	4 kbps to 32 kbps	Scenario Size	1000m x 1000m
Mobility Speed	1-15m/s	Simulation Time	600s
Number of Nodes	50	MAC Protocol	802.11
No of Sources	10	Mobility Model	Random Waypoint
Transmitter Range	250m	Attack Type	Gray-hole, Black-hole
Bandwidth	11Mbps	No of Malicious Nodes	10%
Traffic	CBR over UDP	Protocols	AODV,DSR, DSDV, OLSR

4 Performance Evaluation

Performance analysis of routing protocols in MANET under varying pause time and data rate of the nodes has been performed. Moreover, we have compared and discussed the performance of these well-known MANET routing algorithms under three different scenarios (a) without attack (when there is no security attack on the network), (b) under gray-hole attack, and (c) black-hole attack.

4.1 Packet Delivery Ratio (PDR)

Our simulation results showed that PDR of considered routing protocols under varying pause time (data rate set to 16kbps). AODV and OLSR have highest PDR compared to other two routing protocols. Moreover, the route request and response mechanism of AODV and proactive route maintenance of OLSR also enable these protocols to perform well under high mobility (or low pause time). PDR of DSDV is least among considered protocols because of its complex routing table maintenance at each node. Fig. 1(a) illustrates that PDR of all the routing protocols decrease under gray-hole attack. AODV is more vulnerable to attacks compared to OLSR and DSR. It can also be seen from Fig. 1(b) that under black-hole attack, PDR of all the protocols decrease drastically. Our results show that under attack OLSR performs better in terms of PDR compared to other routing protocols because OLSR has election based routing structure that minimizes the probability of malicious nodes to be the part of active route. AODV and OLSR outperform all other protocols in terms of PDR. DSR performs very well in low data rate, but with an increase in data rate, PDR decreases. Under high data rate, DSR suffers from stale route caching entries at each node that causes data queues to fill up and hence drop of packets. DSDV is not much influenced by the data rate but compared to other routing protocols, it has less PDR. Fig. 2(a) and 2(b) show that OLSR performs better in terms of PDR compared to all other routing protocols under attack. Moreover, Fig. 2(b) also shows that under the black-hole attack, PDR drastically decreases for all considered routing protocols. Reasons for these behaviours are same that have already been discussed in the context of Fig. 1(a) and 1(b).

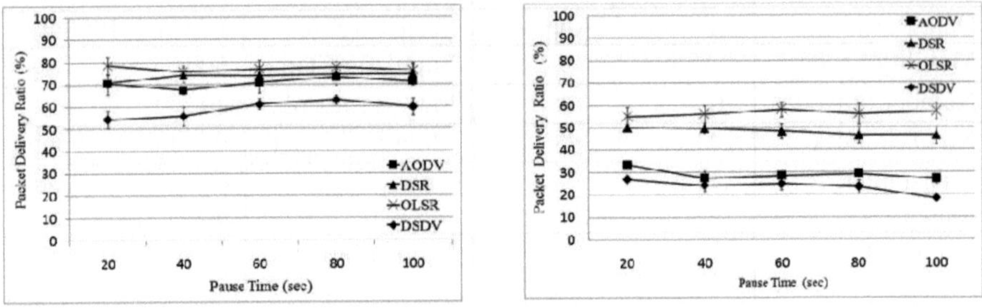

(a) (b)

Fig. 1. (a) PDR vs. pause time under gray-hole attack (b) PDR vs. pause time under black-hole

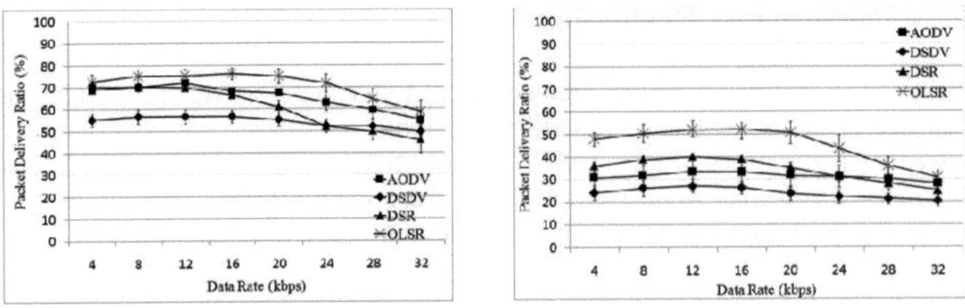

(a) (b)

Fig. 2. (a) PDR vs. data rate under gray-hole attack (b) PDR vs. data rate under black-hole

4.2 Average End-to-End Delay

Fig. 3(a) and 3(b) illustrates that under gray-hole and black-hole attacks, the end-to-end delay of all the routing protocols decrease. This is due to the fact that when the hop count between source and destination pair is large then there is high probability that there exist a malicious node in the path that will drop the packet. As a result, the average end-to-end hop count decreases which eventually leads to reduced end-to-end delay. In the case of gray-hole attack, end-to-end delay is high compared to black-hole attack because black-hole attack causes more packets to drop that result in shorter end-to-end paths and hence smaller end-to-end delay. Moreover, Fig. 3(a) and 3(b) show that DSR has high end-to-end delay because DSR is a reactive protocol, i.e. routing information is not stored in table for each destination in advance. Hence, when data packet arrives at node, it query for path and consumes more time than others. Furthermore, OLSR has high delay compared to AODV and DSDV. In fact it is not the case, actually both AODV and DSDV has low packet delivery ratio compared to OLSR that results in low delay because of the smaller end-to-end hop count that result in reduced end-to-end delay. Since OLSR has higher packet delivery ratio, therefore, average end-to-end distance between source and destination is large that results in high end-to-end delay.

Fig. 4 illustrates average end-to-end delay for varying data rate. It can be observed that DSR has highest average end-to-end delay compared to other routing protocols because of the same reason discussed in the context of Fig 3. Fig. 4 also demonstrates that the average end-to-end delay of all three protocols increase with an increase in data rate. Reason being, high data rate causes congestion that causes an increase in end-to-end delay for all four routing protocols.

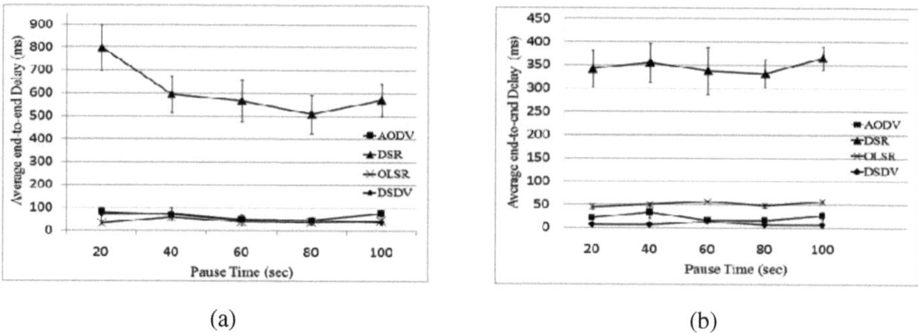

Fig. 3. (a) Average end-to-end delay vs. pause time under gray-hole attack (b) Average end to end delay vs. pause time under black-hole attack

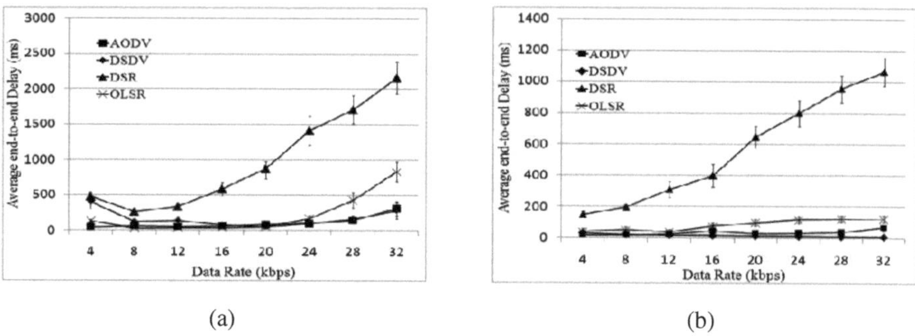

Fig. 4. (a) Average end-to-end delay vs. data rate under gray-hole attack (b) Average end-to-end delay vs. data rate under black-hole attack

4.3 Normalized Routing Load (NRL)

NRL is shown in Fig. 5(a) and 5(b) with varying pause time. Under no attack scenario, NRL of OLSR and DSDV is higher compared to other two routing protocols. This is because of the number of control packets and routing information that is periodically exchange among the nodes and recalculation of MPR nodes particularly when link changes. In DSR, large number of stale route caches at each node and source routing are the cause of high routing load. In AODV, routing load is slightly high because of the inconsistencies in route replies. Under gray-hole and black-hole attack, NRL is less than the case of without attack as shown in Fig. 5(a) and 5(b). This is due to the large number of packets dropped by the malicious nodes and malicious nodes quickly respond to the route requests that reduce the routing traffic in that network. NRL under varying data rate is shown in Fig. 6. NRL of OLSR is very high compared to other protocols in low data rate because less data packets are sent by the nodes against high number of routing packets (proactive nature of the routing protocol). Same is the case with DSDV where NRL decreases with an increase in the data rate. On the other hand in reactive protocols NRL is low for low data rate and increase with an increase in the data rate. Fig. 6(a) and 6(b) depicts that under attacks NRL decreases for all protocols because of the reason discussed in the context of Fig.5.

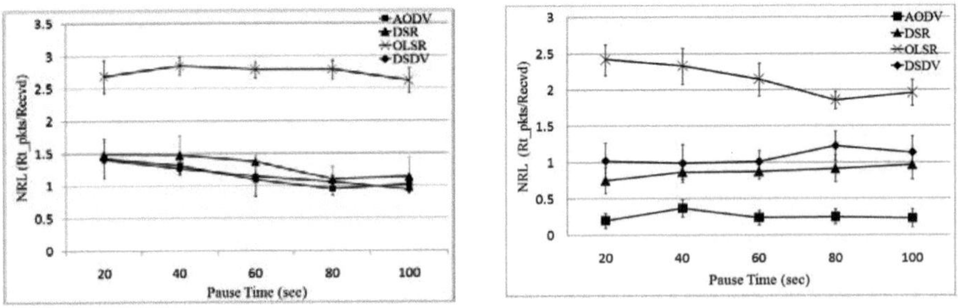

(a) (b)

Fig. 5. (a) NRL vs. pause time under gray-hole attack (b) NRL vs. pause under black-hole attack

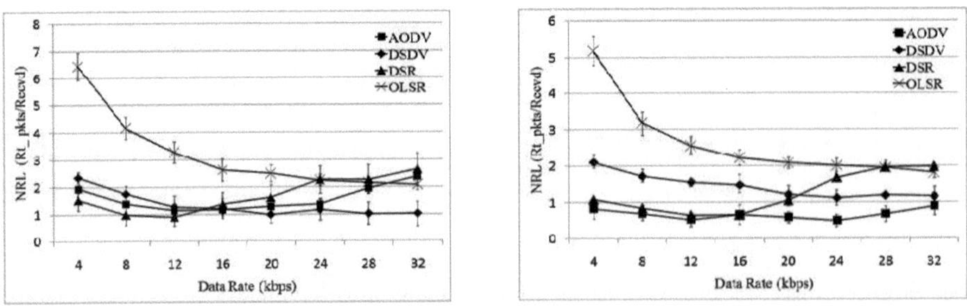

(a) (b)

Fig. 6. (a) NRL vs. data rate under gray-hole attack (b) NRL vs. data rate under black-hole attack

5 Comparative Analysis of MANET Routing Protocols

Table 2 summaries our findings regarding four considered routing protocols (OLSR, DSDV, AODV, and DSR). AODV performs better compared to other routing protocols both in terms of packet delivery ratio under varying mobility and data rate without attacks. However, it has been observed that AODV is more vulnerable to security attacks compared to other routing protocols. DSR performs better compared to other protocols but only under low mobility and low data rate. Under attack scenario, OLSR outperforms other considered routing protocols in terms of PDR with slight higher average end-to-end delay and NRL. Moreover, OLSR performs almost as well as AODV under no attack; however, it has slightly higher NRL compared to AODV. DSDV comparatively has less PDR than other routing protocols. Also it performs poor under high node mobility and high data rate.

6 Related Work

In Ref. [12], performance of reactive and proactive routing protocols was discussed under both active and passive attacks. It was observed that proactive protocols achieve better performance in terms of throughput. Ref. [13] describes the importance of location of malicious nodes in the network. Malicious nodes positioned close to the receiver node are more harmful compared to the case where they are near the sender node. In [14] AODV is analyzed under various attacks. There simulation results reveal that the black hole attack was the most devastating one. Ref. [15] compares and analyzes the performance of AODV and OLSR using ns-2. Both TCP and UDP traffic types were used. AODV performs better with less number of source nodes, whereas OLSR is more effective when there is high number of source nodes. In Ref. [16], the author studied the impact of black hole attack on MANET performance using reachability, hop count, neighbour node density and path optimality metrics. This study reveals that MANET performance degrades significantly with increasing number of black hole nodes. Ref. [17, 18] studied the impact of wormhole attack in proactive MANETs routing protocols. It can be concluded from the study that wormhole degrades the throughput of the network as the number of sources increases.

Table 2. Overall Performance Comparison of MANETs Routing Protocols

Routing Protocols	Attack	PDR	Average end-to-end delay	NRL
AODV	Without	H	H	A
	With	L	A	A
DSR	Without	H	H	H
	With	A	H	A
OLSR	Without	H	A	H
	With	A	A	H
DSDV	Without	A	A	A
	With	L	A	A

7 Conclusion

In this work, a comprehensive performance analysis of the four well-known routing protocols (OLSR, DSDV, AODV, and DSR) has been performed. It has been observed that under normal operation of the network (without attack), each of these well-known protocols have their pros and cons and it is not possible to name one best protocol. However, all of the considered routing protocols were designed without considering the security aspects. As a result, those protocols which were performing best under no attack scenarios fail to give same performance under attacks. This emphasizes the need of handling security aspects during the design of a routing protocol for MANETs. Moreover, there is a need for generic security framework which available protocols can use to minimize the impact of malicious nodes by isolation from the routing path.

References

1. Perkins, C.: Mobile Ad-Hoc Networks. Addison-Wesley (2000)
2. Wieselthier, J., et al.: Wireless ad hoc networks – part II. IEEE Journal on SAC 23(3) (March 2005)
3. Jacquet, P., et al.: Optimized Link State Routing Protocol for Ad Hoc Networks. In: Multi Topic Conference, IEEE INMIC, Technology for the 21st Century, Proceedings, pp. 62–68 (2001)
4. Perkins, C., Bhagwat, P.: Highly Dynamic Destination-Sequenced Distance-Vector (DSDV) routing for mobile computers. Conference on Communications Architectures, Protocols and Applications, SIGCOMM 24(4), 234–244 (1994)
5. Perkins, C., Belding, E., Das, S.: Ad-hoc on-Demand Distance Vector Routing. In: Proceedings of the IEEE Workshop on WMCS, pp. 90–100 (1999)
6. Johnson, D., et al.: The Dynamic Source Routing Protocol for Multihop Wireless Ad Hoc Networks. In: Ad Hoc Networking, pp. 139–172. Addison-Wesley (2001)
7. Abusalah, L., Khokhar, A., Guizani, M.: A Survey of Secure Mobile Ad Hoc Routing Protocols. IEEE Communication Surveys & Tutorials 10(4), 78–93 (2008)
8. Sharma, N., Sharma, A.: The Black-hole Node Attack in MANET. In: Advanced Computing & Communication Technologies (ACCT), pp. 546–550. IEEE (2012)
9. Sadeghi, M., Yahya, S.: Analysis of Wormhole Attack on MANETs using different MANET Routing Protocols. In: ICUFN, pp. 301–303. IEEE
10. Shanmuganathan, V., Anand, T.: A Survey on Gray-hole Attack in MANET. IJCNWC 2(6), 647–650 (2012)
11. Venkataraman, R., Pushpalathe, M., Rao, T.: Performance Analysis of Flooding Attack Prevention Algorithm in MANETs. In: WASET, pp. 458–461 (2009)
12. Sharma, Y., Sharma, A., Sengupta, J.: Performance Evaluation of Mobile Ad hoc Network Routing Protocols under Various Security Attacks. In: ICMCS, IEEE, pp. 117–124 (2010)
13. Palanisamy, V., Annadurai, P., Vijayalakshmi, S.: Impact of Black-hole Attack on Multicast in Ad hoc Network (IBAMA). IEEE Computational Intelligence and Computing Research (2010)
14. Alkatheiri, S., Liu, J., Sangi, R.: AODV Routing Protocol under Several Routing Attacks in MANETs. In: IEEE 13th Conference on Communication Technology (ICCT), pp. 614–618 (2011)

15. Anjum, S., Bhadauria, S.: TCP and UDP based Analysis of AODV and OLSR in Mobile Ad-hoc Networks. In: International Conference on Communication Systems and Network Technologies, pp. 170–173. IEEE (2011)
16. Nagpal, C.K., Kumar, C., Bhushan, B., Gupta, S.: A Study of Black Hole Attack on MANET Performance. International Journal of Modern Education and Computer Science, 47–53 (2012)
17. Ali, S., Wahid, A.: Performance Evaluation of Routing Protocols under wormhole Attack in Mobile Ad Hoc Network. VSRD International Journal of Computer Science & Information Technology 2(5), 394–402 (2012)
18. Patel, V.P.: Reactive and Proactive Routing Protocol Performance Evaluation for Qualitative and Quantitative Analysis in Mobile Ad Hoc Network. International Journal of Scientific and Research Publications 2(9) (September 2012)

Enhanced Speaker Verification Using GMM-Supervector Based Modified Adaptive GMM Training

Tan Dat Trinh[1], Min Kyung Park[1], Jin Young Kim[1],
Kyong Rok Lee[2], and Keeseong Cho[3]

[1] Chonnam National University, Gwangju, Rep. of Korea
[2] Nambu University, Rep. of Korea
[3] ETRI, Daejeon Rep. of Korea
min-0-42@nate.com

Abstract. In this paper, an enhanced speaker verification is proposed by exploring a novel modified adaptive Gaussian mixture model (GMM) training. Based weight factor of observation called the observation reliability; we propose to apply a modified Expectation maximization (EM) algorithm, combined with a modified Maximum a posteriori (MAP) estimation to train the modified adaptive GMM model. Using this proposed model, we generate GMM-supervectors which are combined with SVM for verification task. We evaluate performance of speaker verification system based the proposed approaches on utterances from Korean movie database ("You came from the stars"). Experimental results demonstrate that our proposed approaches can outperform the standard GMM-UBM and GMM-supervector approaches in noise conditions.

1 Introduction

Recently, text-independent speaker verification in noise conditions and channel variations becomes one of the most interesting challenges. Many different approaches have been proposed to deal with these issues. Joint factor analysis (JFA) [1] has been applied successfully to speaker verification to compensate speaker and channel variations in the context of Gaussian mixture model-Universal background model (GMM-UBM) [2]. In addition, GMM-supervector concept has been discovered by stacking means of the GMM-UBM mixture components using Maximum a posteriori (MAP) adaptation. The combination between SVM and GMM-supervector has been proven to obtain interesting results. Nuisance attribute projection (NAP) was applied to reduce the channel effect [3]. In [4], JFA and SVM using cosine kernel gave the most significant results.

In real applications, utterances are often distorted and corrupted by additive or convolutive noise, and channel effect. These signal observations can reduce the speaker verification performance. An improved method based observation reliability in the GMM model has been successfully applied to speaker recognition [5].

K.J. Kim and N. Wattanapongsakorn (eds.), *Mobile and Wireless Technology 2015*,
Lecture Notes in Electrical Engineering 310, DOI: 10.1007/978-3-662-47669-7_17

Our main aim in this paper is to investigate the effect of the observation reliability in an adaptive GMM algorithm, and its application to speaker verification under noise conditions. Based the observation reliability, we propose a modified version of adaptive GMM model trained via the modified Expectation maximization (EM) algorithm, combined with the modified Maximum a posteriori (MAP) estimation. Frame-based segmental SNR value is used as a measure of the observation reliability. We also propose to construct GMM-supervectors based the modified model. An SVM with a non-linear GMM KL kernel function are used to evaluate the performance of verification system.

2 Enhanced Speaker Verification System Based Observation Reliability

2.1 Modified Adaptive GMM Model

In real speaker verification application, the performance of verification system can be degraded under noise conditions. Specially, additive or convolutive noise creates a mismatch between training and testing phases. In our experiments, different types of noise influence speech such as music effect, traffic noise, cell phone ringing sound effect, sob sound effect, and so on. Most of the works previously in speech/speaker recognition attempt to recover clean speech from noisy speech. Many methods have been proposed to reduce noise and channel variations. However, they cannot be completely discarded.

In the classical adaptive GMM training [2], observation vectors can be considered generated from noise-free speeches. They are evenly treated equally without any attention of their reliability. However, in real application, the observation vectors are usually affected and corrupted by noise. These affected parts often obtain low reliability, and can reduce the speaker verification performance. It is reasonable to investigate the observation reliability in the adaptive GMM training. Based on the observation reliability, we propose to modify the adaptive GMM training to estimate speaker models.

We assume that each observation has a relevant reliability value. Specially, each observation vector can be considered as possibly having a different reliability value, and treated differently. For example, the segmental SNR of the speech signal may vary in each frame. A modified version of the EM algorithm based on the observation reliability to estimate GMM parameters was proposed in [5]. In addition, the GMM parameters may also be estimated using Maximum a posteriori (MAP) estimation. MAP estimation is used to derive a speaker model by adapting from a Universal background model (UBM) [2]. In this study, to obtain the UBM model, we simply pool all the background data, and use it to train the UBM via the modified EM algorithm. Using the UBM model, we propose a modified version of MAP estimation based the observation reliability to estimate the GMM parameters.

The specifics of the modified adaptation are as follow. Given a UBM model estimated using the modified EM algorithm $\theta_{UBM} = \left(\pi_i^u, \mu_i^u, \Sigma_i^u\right), i = 1, \ldots, M$, where M is number of mixtures and training vectors from a desired class for each speaker $Y = \left\{y_n\right\}_{n=1}^{N}$. Assume that $f\left(y_n \mid \mu_i^u, \Sigma_i^u\right)$ is a D-variate Gaussian function. For mixture i in the UBM model, we compute a posteriori probability as:

$$\Pr\left(i \mid y_n, \pi_i^u, \mu_i^u, \Sigma_i^u\right) = \frac{\pi_i^u f\left(y_n \mid \mu_i^u, \Sigma_i^u\right)}{\sum_{j=1}^{M} \pi_j^u f\left(y_n \mid \mu_j^u, \Sigma_j^u\right)}. \quad (1)$$

Then modified sufficient statistics for the weight, mean and variance parameters are computed:

$$n_i = \sum_{n=1}^{N} \alpha_n \Pr\left(i \mid y_n, \pi_i^u, \mu_i^u, \Sigma_i^u\right). \quad (2)$$

$$E_i\left(y\right) = \frac{\sum_{n=1}^{N} \alpha_n y_n \Pr\left(i \mid y_n, \pi_i^u, \mu_i^u, \Sigma_i^u\right)}{n_i}. \quad (3)$$

$$E_i\left(y^2\right) = \frac{\sum_{n=1}^{N} \alpha_n y_n^2 \Pr\left(i \mid y_n, \pi_i^u, \mu_i^u, \Sigma_i^u\right)}{n_i}. \quad (4)$$

where α_n is the observation reliability value.

Eventually, using these new sufficient statistics from the training data, we can update the old UBM sufficient statistics for mixture i to form the adapted parameters for the mixture i with the equations described in [2]. In the above equations, the modified MAP estimation is actually very similar to that of the baseline MAP estimation. If the observation reliability values are actually equal to 1, the modified algorithm is exactly the same as the baseline MAP estimation.

To estimate the modified adaptive GMM model, we first need to calculate the observation reliability value. In this study, the reliability value of observation is computed from signal-to-noise ratio (SNR). We can assume that the input speech is affected by noise. The optimal minimum mean square error (MMSE) short-time spectral amplitude (STSA) technique is used to estimate an enhanced speech from noisy speech. This method has proven effective at significantly reducing the problem of music noise by recursively smoothing the a priori SNR [6]. We use MMSE log-STSA with VAD-based noise estimate to find the enhanced speech. Segmental SNR is then estimated between the input speech and the enhanced speech. These functions were implemented using the VOICEBOX toolbox [7]. An example of segmental SNR between the input speech and enhanced speech (reference speech) is shown in Fig. 1.

In our experiments, the segmental SNR is first computed, and it is then transform into the observation reliability value ranging from *0* and *1* using a simple sigmoid function.

$$\alpha_n = g\left(SNR_n, s, c\right) = \frac{1}{1 + e^{-s(SNR_n - c)}} . \quad (5)$$

where s is a scaling parameter, c is a shift parameter, and SNR_n is the estimated SNR of *n-th* observation. In this paper, c is mean of segmental SNR values of an input signal and s is chosen of 0.55. Fig. 2 shows an example of the observation reliability with the parameters $c = 3.07$ and $s = 0.55$.

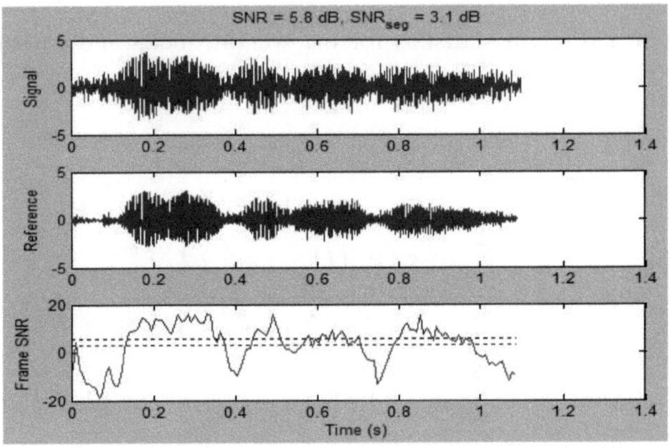

Fig. 1. Example of segmental SNR between the input speech and reference speech

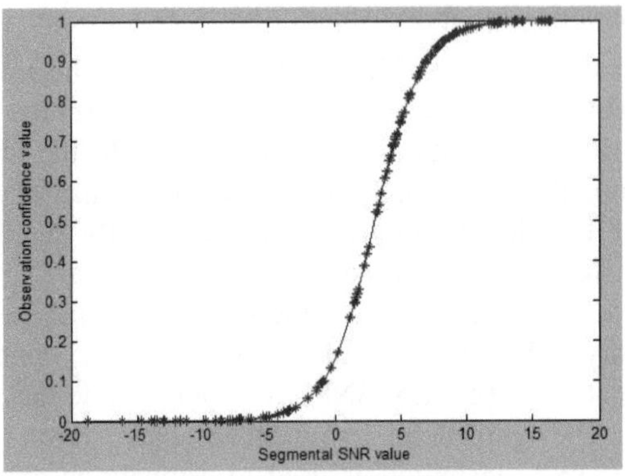

Fig. 2. Observation reliability based on sigmoid function

2.2 GMM-Supervector Based Observation Reliability

In this section, we propose to form GMM-supervectors based on the modified adaptive GMM training and observation reliability presented in the previous section. We adapt both the Gaussian mean and covariance information for the SVM kernel.

Fig. 3 shows a schematic diagram of constructing the modified GMM-supervector based the observation reliability. In the diagram, solid lines describe the standard GMM-supervector system, and dashed lines describe our improved processes added to the standard method. We use a non-linear GMM KL kernel function [8] to evaluate the performance of SVM based speaker verification system.

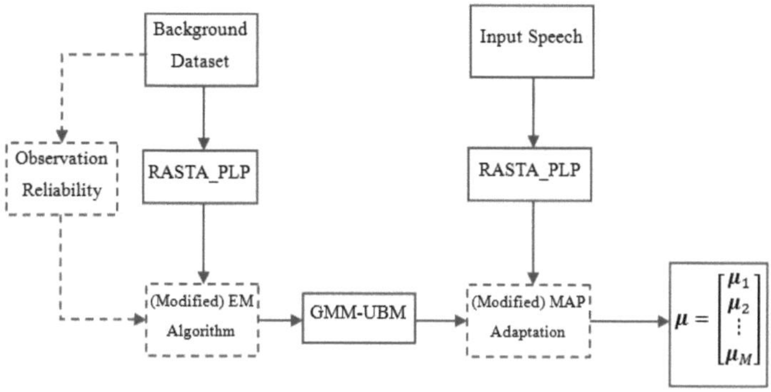

Fig. 3. Schematic diagram of forming GMM-supervector based the observation reliability.

3 Experiments and Results

Korean movie database ("You came from the stars") is used to evaluate performance of speaker verification system. Audio event samples in the first twelve episodes were manually segmented and clustered based corresponding events. We selected seven target speakers to investigate the performance of verification system. The samples had different length (0.5s to 10s) and were normalized with 16 bit-precision at a sampling rate of 16 kHz. Training set was constructed from all clean speeches of each target speaker extracted from the first six episodes. We divided testing set into two groups: Clean Test and Noisy Test. All clean speeches of each target speaker extracted from the rest six episodes and 1200 random samples from other events were constructed as Clean Test. We gathered all noise-corrupted speeches (such as music effect, traffic noise, cell phone ringing sound, sob sound, and so on) of each target speaker extracted from the rest six episodes and 1200 random samples from other events to create Noisy Test. Information of training and testing dataset is shown in Table 1.

To reduce effect of noise in utterance, we apply Independent component analysis (ICA) [9] to two channels of input utterance to find enhanced speech source. Then, RASTA_PLP [10] is used to extract feature vectors from speech source in ICA's output. An 42-dimensional cepstral feature vector consisting of 13th order PLP features with RASTA combined with delta and delta-delta coefficients is computed. The features are calculated every 8ms using a 16 ms Hanning window. We also apply feature mapping technique to remove channel effects [11].

Table 1. Example of audio samples used for experiments

Speakers	Number of Samples		
	Training	Testing	
		Clean Test	Noisy Test
Speaker 1	**170**	**208**	**611**
Speaker 2	**348**	**513**	**1167**
Speaker 3	**44**	**114**	**179**
Speaker 4	**162**	**137**	**326**
Speaker 5	**53**	**22**	**216**
Speaker 6	**57**	**61**	**33**
Speaker 7	**11**	**87**	**111**

To compose the GMM-UBM (impostor model), 128 mixture components are chosen. It leads to 5376-dimension of the GMM-supervector. The GMM-UBM is estimated using the proposed adaptive GMM training algorithm, where a relevance factor is set of 16, from 3000 samples (including 1000 random samples from target speakers, and 2000 random samples from other speakers and events). A background SVM consists of GMM-supervectors labeled as -1 extracted from 2000 random samples from other speakers and events. Then an SVM model is trained using the target GMM-supervectors and the SVM background. LIBSVM version 3.13 [12] is used evaluate the performance of system. In test stage, the test utterances are first identified, and then we compare calculated scores to a threshold to make the final decision. Equal error rate (EER) is used as metrics for evaluation.

Table 2. Comparison between different systems on our dataset, in terms of equal error rate (EER)

System	EER(%)	
	Clean Test	Noisy Test
Standard GMM-UBM	11.31	16.57
Modified GMM-UBM	**10.71**	**12.45**
Standard GSV-SVM	13.89	16.94
Modified GSV-SVM	**13.34**	**15.21**

We compare the performance between the proposed approaches with the standard well-known approaches. Table 2 gives the experimental results of the speaker verification system. In this table, two proposed approaches, described as modified GMM-UBM and modified GSV-SVM, are GMM-UBM and GMM-supervector based SVM systems using the modified adaptive GMM training, respectively. Experimental results revealed that the proposed approaches outperform the standard systems in noise conditions. The proposed approaches have demonstrated promise for enhancing the performance of the speaker verification system.

4 Conclusion

In this paper, an enhanced speaker verification system by modifying adaptive GMM training based the observation reliability value has been proposed. In addition, we have introduced to construct GMM-supervector using the reliability value. We have applied SVM with the non-linear GMM KL kernel function to evaluate the performance of speaker verification system. The proposed approaches are effective and robust under noise conditions. The experimental results have shown that our proposed approaches can significantly enhance the performance of the verification system, and outperform the baseline approaches.

Acknowledgments. This work was supported by the ICT R&D program of MSIP/IITP. [14-000-11-002, Development of Object-based Knowledge Convergence Service Platform using Image Recognition in Broadcasting Contents]

References

1. Kenny, P., Boulianne, G., Ouellet, P., Dumouchel, P.: Joint Factor Analysis versus Eigenchannels in Speaker Recognition. IEEE Trans. Audio, Speech, Lang. Process. 15(4), 1435–1447 (2007)
2. Reynolds, D.A., Quatieri, T.F., Dunn, R.B.: Speaker Verification using Adapted Gaussian Mixture Models. Digital Signal Processing (10), 19–41 (2000)
3. Campbell, W.M., Sturim, D.E., Reynolds, D.A., Solomonoff, A.: SVM based Speaker Verification using a GMM Supervector Kernel and NAP Variability Compensation. In: Proc. IEEE ICASSP, vol. 1, pp. 97–100 (2006)
4. Dehak, N., Kenny, P., Dehak, R., Glembek, O., Dumouchel, P., Burget, L., Hubeika, V., Castaldo, F.: Support Vector Machines and Joint Factor Analysis for Speaker Verification. In: Proc. IEEE ICASSP, pp. 4237–4240 (2009)
5. Kim, J.Y., Min, S.H., Na, S.Y., Choi, H.S., Choi, S.H.: Modified GMM Training for Inexact Observation and Its Application to Speaker Identification. Speech Sciences 14, 163–175 (2007)
6. May, T., Par, S.V.D., Kohlrausch, A.: Noise-Robust Speaker Recognition Combining Missing Data Techniques and Universal Background Modeling. IEEE Trans. Audio, Speech, Lang. Process. 20(1), 108–121 (2012)
7. Brookes, M.: Voicebox: Speech Processing Toolbox for Matlab (2007), http://www.ee.ic.ac.uk/hp/staff/dmb/voicebox/voicebox.html
8. Bui, N.N., Kim, J.Y., Trinh, T.D.: A Non-Linear GMM KL and GUMI Kernel for SVM Using GMM-UBM Supervector in Home Acoustic Event Classification. IEICE Trans. Fundamentals E97-A(8), 1791–1794 (2014)
9. Senturk, A., Gurgen, F.S.: Feature Selection by Independent Component Analysis for Robust Speaker Verification. International Journal of Computer Science and Network Security 6(3B), 229–239 (2006)
10. Hermansky, H., Morgan, N.: RASTA Processing of Speech. IEEE Trans. on Speech and Audio Proc. 2(4), 578–589 (1994)
11. Pelecanos, J., Sridharan, S.: Feature Warping for Robust Speaker Verification. In: Proc. Speaker Odyssey, Crete, Greece, pp. 213–218 (2001)
12. Hsu, C.W., Chang, C.C., Lin, C.J.: A Practical Guide to Support Vector Classification (2010), http://www.csie.ntu.edu.tw/~cjlin/libsvm

Fit Buddy: A Mobile Application for Fitness Tracking Using the Always-On Low-Power Sensor

Kasidit Wijitsopon, Chavalit Panichayanubal, and Pusadee Seresangtakul

Department of Computer Science, Khon Kaen University,
123 Mitraphab Road Nai-Meuang Meuang, Khon Kaen Thailand
kasidit@softsion.com, ghostrest@hotmail.com, pusadee@kku.ac.th

Abstract. This paper presents an always-on and always accessible mobile application, namely Fit Buddy, which will help users track their personal fitness statistics. The application focuses on step counting from both walking and running using a smartphone. The Fit Buddy can be linked to a user's social media account and rank the user's fitness data compared with those of friends who are on the same social network. By utilizing SensorCore data from a smartphone, user's daily steps, active time and location can be tracked and recorded without the need for human interaction because of the always-on aspect of the application. Users also have immediate access to the history data. The Fit Buddy makes personalized daily goals easier to commit and achieve. It gives the users a simple and effective method of motivation to do more exercise.

1 Introduction

Presently, the modern society is extremely influential in human lives; consequently, their behavior has been changed. In particular, the way of human life style has rapidly involved with advanced technologies. Moreover, with the new life style, they have to rush to complete their daily routine. By the way, some of their duties require them to sit constantly at work, resulting in a decrease in body movement. In addition, people ignore to pay more attention to their health appropriately [1]. This leads not only to body degradation but also to the causes of the physical abnormalities and various diseases.

In medical context, exercising is the best way to protect against and cure physical deterioration, and it can rehabilitate those physical problems. According to the medical problems, such as diabetes, high-blood pressure, coronary heart disease, hemorrhagic stroke, and chronic obstructive pulmonary disease, they are caused by the chronic illnesses, such as obesity. In other words, it is Metabolic Syndrome condition [2].

In present, there are many wearable devices being made by many brands. People can wear these devices, track their activities, and then sync the data to a smartphone. However, they have to buy the devices separately. That is, a smartphone doesn't come bundled with the devices [5],[6],[7],[8]. Moreover, developers have to develop

K.J. Kim and N. Wattanapongsakorn (eds.), *Mobile and Wireless Technology 2015*,
Lecture Notes in Electrical Engineering 310, DOI: 10.1007/978-3-662-47669-7_18

smartphone applications to motivate people to do exercise [10],[11]. Currently, a smartphone is widely popular, and the new generation of a smartphone offers a new built-in sensor, named SensorCore, which can track our daily life activity without buying the wearable devices.

Consequently, the need and motivation discussed above are the reason why we developed a new smartphone application. In other words, we aim to encourage people to realize that exercising is the best choice to stay healthy. Moreover, this app can observe the behavior of those people who want more data about their activities and their health-related statistics which are always-on, and always accessible. Beneficially, the app can develop an awareness of their health condition and can help them set their target of burning energy routinely. Therefore, its high efficiency in ranking the burned energy is the key feature. Therefore, the users can rank and compete with friends to be the person who burned the most calories. This app can also motivate people to concern about, and keep exercising as much as they can.

The main purpose of this paper is to develop an application on Windows Phone 8.1 that is used for monitoring people's daily activity to track walking and running steps by using SensorCore. In addition, the application also provides the calculation of calories burned from daily life activities.

2 Related Work

Microsoft Mobile has developed Lumia SensorCore SDK [4] that embeds along with the new release smartphones, such as Nokia Lumia 930, Nokia Lumia 630 and Nokia Lumia 1520. The SensorCore has functions of storing the data about the users' movement, whether walk or run. This gives a chance for the developers to develop the applications. The SensorCore is working behind the background process throughout the smartphone which is always-on, without impact on the main memory and battery of the smartphone. The data about various movements, whether walk or run, perhaps the users are standing, are always stored into the memory which limits data from SensorCore into the main memory of the smartphone and it can be able to check backward after only 10 days. Several developers [5],[6],[7],[8] have developed the gadgets that detect users' health by using wearable devices: the wristband or GPS watch. These devices keep detecting and following health activities data daily. The users always wear it whether walk, run or sleeping. It can tell about calorie burned in a day. Those wearable devices are going to collaborate with an application on the users' smartphone, which can share those data on social network or race with friends by leaderboard to rank a score, which make the users motivate themselves to exercise more than usual. However, as we mentioned before, users have to buy the wearable devices. RunKeeper [9] is an efficient application for running workouts, which is able to calculate running distance and the number of energy used by running, acceleration and time spent during the break. Users can set their goal to exercise by using the data from GPS to calculate distance and speed of their running. When the users finished their running, the program can record their activities statistical data, and view their previous data. This application can sync to social network to increase the number of friends, post photo and message, and share information with their friends. In order to track their activities, users must start and stop tracking manually when they start or finish their running. Z. He and Y. Luo researched and developed Runking [10] application on Android background process, which used for exercise. The Runking is able to rank a number of the users from the statistical data of running; thus, this creates a competition. The program is activated automatically whenever users start their activities. The Runking requires collaboration from the sensor Accelerometer of the smartphones that can learn the characteristic and behavioral activities of the users. The Runking can use the data from GPS to detect distance and the period of time that users used. However, the Runking must be running as a background task. R. de Oliveira, M. Cherubini and N. Oliver researched and developed Movipill [11], a social game to motivate the users who are elderly and to inform them to take medicine quantity correctly and punctually every day, especially, the elders who have problems with memory. The users have to take medicine punctually as prescribed. NajmeZehra Naqvi, Ashwani Kumar, Aanchal Chauhan and KritkaSahni proposed the algorithm on counting steps with accelerometer [12] by using the force in z axis to analyze the step signal data, with a frequency of 100Hz. The researchers examined the result by placing the smartphone in the position on the waist of the users and let testers standing for 5 a seconds before stared to evaluate the threshold of stand static. They showed the result of the number of steps in graph. Microsoft has developed the Cloud Platform [13] that can empower applications and various services to scale. It is able to adjust itself according to the users' need. PaaS (Platform as a Service) and laaS (Infrastructure as a service) are cloud computing service. It provides a platform and environment for developers to develop, manage, and run their applications, without the complexity of building. Moreover, Microsoft developed WNS (window notification service) [14] as one of services, that can help developers to send information in a kind of Toast, Tile, Badge and Raw from the cloud service to the users. The WNS has improved the mechanical working, more capability, and certify working on a kind of application Windows Runtime. Facebook has developed the Graph API [15] for developers to connect and access the data on Facebook service. Developers can develop applications directly connected to Facebook in order to share photo or get private data from the list of friends. In addition, the accessibility has to be permitted by the users before. The Graph API is created along with the REST architecture [16]. It is one of the architectures used for data communication in feature of JavaScript Object Notation (JSON) [17]. It is one of ways for exchanging data in a feature message that has been developed from ECMA Script [18]. It is used to substitute the kind of the main messages in the UTF-8 representation. In addition, the JSON data type can be exchanged without depending on the language of the program, and also can be identified to be the standard of RFC4627. Apple has developed the iOS 8 [20] which is certificated an access health data of the users, which can count walking steps that can find on the new smartphones, the latest one to the iPhone 5s, because it must have the M7 Co-Processor, which was embed along with the new generation smartphones of Apple. It is responsible for processing the health data, in particular.

3 System and Architecture

The system comprises 2 parts: client and server parts. The client was developed on Windows Phone 8.1 and the server application was deployed on Microsoft Azure, which is cloud platform (Fig. 1 shows the detail).

The Fit Buddy Cloud Service consists of 4 main parts, which are Fit Buddy API, Data Management, Windows Push Notification, and Graph API.

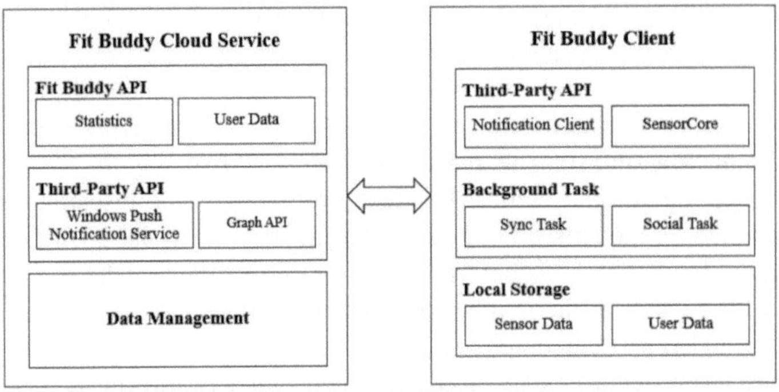

Fig. 1. Shows the Fit Buddy architecture

The Fit Buddy API is developed by using PHP, high efficiency server-side programming languages, flexible for database connectivity and fast load time. It will sync together with Fit Buddy Client. The Fit Buddy API is the web service, which is created according to the REST [16] architecture. It works as a mediator between server and client. The Fit buddy API will provide services to support information requests as well as information submissions from the Fit Buddy Client. It will respond the requests of the information in JSON format. The examples of requested information include statistical data, users' profile and ranking.

Concerning Data Management, this module provides functions to control the data accessibility and secure the database. In this work, we choose MySQL to be the database server because it is an open source and it is the most widely and popular DBMS. It is highly effective that can scale and work together with PHP.

In Windows Push Notification, this module is directly connected to Windows Push Notification Services [14], which enables the Fit Buddy Cloud Service to update the data, including summation of the toast, tile or raw data formats, and send to the Fit Buddy Client instantly. The Fit Buddy Client is registered with push notification channel. It will send URI to Fit Buddy Cloud Services.

Regarding Graph API, this module uses Facebook login, which enables the Fit Buddy to connect to the users' data from Facebook, such as full name, profile picture, and the date of birth. Moreover, it can share photos and the fitness data of the users on Facebook. The Fit Buddy Client consists of 4 main modules, including Local Storage, Background Task, Notification Client and SensorCore. The details are shown as followings.

Concerning Local Storage Management, this module is used to manage application storage. The application will keep track of user general data during 10 days. The history data over 10 days will be stored in local storage which is managed by the local storage management module. The stored data is general information which contains user name, last name, birth date, weight, height and daily energy usage goal.

In Background Task, there are 2 background tasks, including Sync and Social Tasks. The Sync Task is processing all the time. The Sync task will send and store the sensor data into the Fit Buddy Cloud Service in every 30 minutes. The Social Task can collect the social data from the Fit Buddy Cloud Service, such as users' ranking, it will compare the ranking with the ranking of friends who use the Fit Buddy.

In terms of Notification Client, this module is used to receive the data sent from the Fit Buddy Cloud Service to the users. In order to receive the data, this module must be registered with the Windows Push Notification Service.

Talking about SensorCore, this module collects motion data which consists of activity type, the number of walking and running steps from the mobile sensor core. The sensor can tap into 10 days. This part always works as the background. This makes the users interact with the application anytime.

4 Implementation and Evaluation

The system and architecture were developed on Windows Phone 8.1 operating system. We implemented and tested on Nokia Lumia 630, a low-end smartphone sold in the market. The Fit Buddy user interface shows in Figure 2 and Figure 3.

Fig. 2. Shows the example of front screen of the Fit Buddy user interface

The accuracy test of walking and running steps that was counted by SensorCore was tested by10 testers. Each person walked and run 100 steps by placing the smartphone on different parts of the testers' body.

From Table 1, we concluded the step count by asking the testers to walk 100 steps and placed the smartphone on different body parts. The app could get accurate number of walking steps by placing the smartphone on the waist, in hand and in the pocket consecutively. By running 100 steps, the most accurate numbers were obtained by placing the smartphone on the waist and on the arm consecutively.

Table 1. During the accuracy test of walking steps, the smartphone was placed on a waist, in a hand and in a pocket. Concerning the accuracy test of step count by running 100 steps, the smartphone was belted on a waist, carried in hand, and strapped it on the arm.

Sample	Walk			Run		
	Waist	Hand	Bag	Waist	Arm	Hand
1	103	95	108	78	78	82
2	101	68	99	94	66	85
3	104	72	104	80	80	85
4	102	64	102	88	89	86
5	101	54	100	76	83	88
6	103	93	102	75	81	84
7	101	71	105	74	77	80
8	100	66	104	79	80	89
9	100	90	102	75	66	87
10	109	67	101	93	69	89
Accuracy	97.60%	74.00%	97.30%	81.20%	76.90%	85.50%

The burned calories are calculated by steps counted from walking, running and upstairs walking. For a walking on 0% of walking surface grade [21], the equation (1) is as following:

$$CB = (0.0215 \times KPH^3 - 0.1765 \times KPH^2 \times 0.8710 \times KPH + 1.4577) \times Weight \times Time \qquad (1)$$

For a running, the burned calories are calculated from a distance and a weight as equation (2) below

$$CB = KM \times Weight \times 1.036 \qquad (2)$$

CB stands for Calories Burned, KPH is Kilometer per hour, Weight is user's mass weight and KM is a distance in Kilometer.

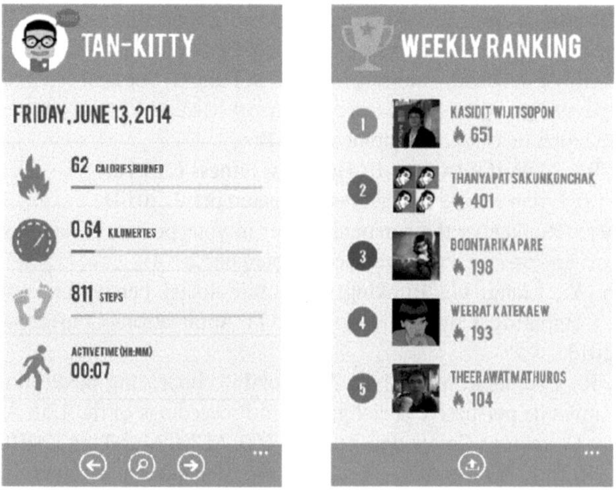

Fig. 3. History and Ranking Screen of Fit Buddy

5 Conclusion and Future Work

This paper presents a mobile application called Fit Buddy for activity monitoring. In the future, we purpose the developers should develop the upstairs walking step count algorithm and detection algorithm for a smartphone positioning.

Acknowledgments. The authors would like to thank Computer Science Department, and Graduate School, Khon Kaen University for funding. We very much appreciate the Developer Experience team from Microsoft Mobile (Thailand) for technical and device supported.

References

1. Maopech, K., Choupanich, K., Lapho, P.: Behaviors for Exercises of Personnel in Kasetsart University, Kampheangsaen Campus. Master Thesis, the Graduate school of KU and KU library, Kasetsard University, Thailand (2012)
2. Alberti, K.G.M.M., Zimmet, P., Shaw, J.: Metabolic syndrome—a new world-wide definition. A consensus statement from the international diabetes federation. Diabetic Medicine 23(5), 469–480 (2006)
3. International Data Corporation (IDC), "Smartphone Momentum Still Evident with Shipments Expected to Reach 1.2 Billion in 2014 and Growing 23.1% Over 2013, According to IDC," IDC Worldwide Mobile Phone Tracker, May 28 (2014), `http://www.idc.com/getdoc.jsp?containerId=prUS24857114` (accessed: August 2014)
4. Microsoft Mobile Oy, "Lumia SensorCore SDK," Nokia Developer Beta (2014), `http://developer.nokia.com/lumia/nokia-apis/sensorcore` (accessed: August 2014)
5. Fitbit Inc., "Fitbit apps:iOS/Android," Health&Fitness (2014), `http://www.fitbit.com/apps` (accessed: September 2014)

6. Runtastic, "Runtastic app, iOS/Android/Windows Phone/BlackBerry," Health & Fitness (2014), http://www.runtastic.com (accessed: September 2014)
7. Nike Inc., "Nike+ FuelBand, Android," Health & Fitness, September 23 (2014), https://play.google.com/store/apps/details?id=com.nike.fb&hl=en (accessed: September 2014)
8. Jawbone's UP®, "UP, iOS/Android," Health & Fitness (2014), https://jawbone.com/up (accessed: September 2, 2014)
9. FitnessKeeper, "Runkeeper:the personal trainer in your pocket," Health & Fitness (2014), http://runkeeper.com/ (accessed: September 2014)
10. He, Z., Luo, Y., Liang, G.: Runking: A mobile social persuasion system for running exercise. In: Computing, Communications and IT Applications Conference (ComComAp), pp. 74–78 (2013)
11. De Oliveira, R., Cherubini, M., Oliver, N.: MoviPill: improving medication compliance for elders using a mobile persuasive social game. In: Proceedings of the 12th ACM International Conference on Ubiquitous Computing, pp. 251–260. ACM, New York (2010)
12. Naqvi, N.Z., Kumar, A., Chauhan, A., Sahni, K.: Step Counting Using Smartphone-Based Accelerometer. International Journal on Computer Science and Engineering (IJCSE) 4(5), 675–681 (2012)
13. Microsoft, "Microsoft Cloud platform," Microsoft Azure, February 1 (2010), https://azure.microsoft.com (accessed: September 2014)
14. Microsoft, "Windows Push Notification Services (WNS) overview (Windows runtime app)," Develop, August 19 (2014), http://msdn.microsoft.com/en-us/library/windows/apps/hh913756.aspx (accessed: October 2014)
15. Facebook Inc., "Graph API Documents website," Developers, May 27 (2007), https://developers.facebook.com/docs/graph-api (accessed: October 2014)
16. Fielding, R.T.: Architectural styles and the design of network-based software architectures. PhD diss., University of California, Irvine (2000)
17. Crockford, D.: The application/json media type for javascript object notation (json). The Internet Society (July 2006), https://tools.ietf.org/html/rfc4627 (accessed: October 2014)
18. European Computer Manufacturers Association, "ECMAScript Language Specification 3rd Edition" (December 1999), http://www.ecma-international.org/publications/files/ecma-st/ECMA-262.pdf (accessed: October 2014)
19. Apple Inc., "App Developers are already doing amazing things with IOS 8," ios8, September 9 (2014), https://www.apple.com/ios/developer (accessed: October 2014)
20. Teh, K.C., Aziz, A.R.: Heart rate, oxygen uptake, and energy cost of ascending and descending the stairs. Medicine and Science in Sports and Exercise 34(4), 695–699 (2002)
21. Margaria, R.: Sulla fisiologia e specialmente sul consumo energetico della marcia e della corsa a varie velocità ed inclinazioni del terreno. Atti Accad Naz Lincei Memorie 7, 299–368 (1938)
22. Öberg, T., Karsznia, A., Öberg, K.: Basic gait parameters: reference data for normal subjects, 10-79 years of age. Journal of Rehabilitation Research and Development 30(2), 210 (1993)
23. Hatano, Y.: Use of the pedometer for promoting daily walking exercise. International Council for Health, Physical Education, and Recreation 29(4), 4–8 (1993)

Firenzina: Porting a Chess Engine to Android

Corey Abshire and Dmitri Gusev

Indiana University, Purdue University
Columbus, IN, U.S.A., Columbus, IN, U.S.A.
dcabshir@indiana.edu, dgusev@purdue.edu

Abstract. We have ported an open-source chess engine to Android using the Native Development Kit (NDK), tested it against other engines, and played against it ourselves. The porting process revealed a number of challenges and insights, some of which may be similar across other mobile application ports, while others are likely unique to chess. We found that few changes were needed to have a working engine, and that the architecture of chess engines based on the universal chess interface (UCI) protocol enabled quick adoption of an advanced user interface. We also found that the native development experience made it difficult to find the path to a complete solution requiring relatively minor changes. In this paper, we present our solutions, along with observations about the porting process in the context of the UCI-based chess engine architecture. We believe that these observations will be of interest to developers of other applications as well.

1 Introduction

Mobile platforms have developed their own software ecosystems that focus on solving problems unique to their environments. Using the provided tools, developers have created numerous applications to run on those platforms, often from scratch within this new ecosystem. However, a large body of software for other platforms, such as desktops and servers, already exists.

Android, the mobile operating system by Google, is based on a Java platform with a customized runtime called Dalvik, which is built on top of Linux. Because of this, a more traditional language, operating system call, and a library interface are available for developers. That makes the development process for Android similar to those for desktop and server platforms. One of the benefits of this is that a relatively straightforward path for porting existing software to the Android platform is enabled. Furthermore, Google has published its Native Development Toolkit (NDK)[1] to allow developers to take advantage of this possibility in an officially supported way.

This paper describes a port of one such piece of software, an open-source, cross-platform chess engine called Firenzina. Available on Github[2], Firenzina, by Schmidt,

[1] https://developer.android.com/tools/sdk/ndk/index.html
[2] https://github.com/Censor/Firenzina

K.J. Kim and N. Wattanapongsakorn (eds.), *Mobile and Wireless Technology 2015*,
Lecture Notes in Electrical Engineering 310, DOI: 10.1007/978-3-662-47669-7_19

Gusev, and Brades, could already be built successfully for Windows and Linux platforms. Like its predecessors, Firenzina is a universal chess interface (UCI) based chess engine. UCI, by Huber and Meyer-Kahlen, is a popular open standard protocol that chess engines and chess user interfaces can use to communicate with each other (Meyer-Kahlen, 2000). Our goal with the port and with this paper was to explore some of the challenges of porting a complex piece of software to a mobile platform, without necessarily getting into the extra complexity of the changes required to meet the interaction design and user experience needs.

The architecture of UCI-based chess engines was a key enabler to this goal. Because it cleanly separates the user interface from the game engine, and because others have ported or created the additional components required to provide the user interface and configuration tools, we were able to focus on the task of porting the chess engine itself. In particular, for this project we made use of Chess for All[3], a UCI-compatible chess GUI for Android by Schreiner (2013), and Chess Engines[4], a configuration and wrapper tool by Bik (2013) to enable ARM-compiled UCI chess engines to be compatible with Android chess GUI's[5] such as Chess for All. Both are available on the Google Play Store[6], free of charge.

With the Android NDK, a cookbook by Liu (2013), a guidebook by Ratabouil (2012), and the set of tools described above, we were able to get Firenzina to run on Android with surprisingly few changes. This paper lists the specific changes required, along with instructions for running Firenzina on your phone using Chess for All and Chess Engines. It also discusses some of our observations around UCI chess system architecture and how it may be applicable to non-chess domains to enable a better porting model.

Even though many valuable sources exist on other aspects of chess programming, notably Levy (1990) and Hyatt (2005), the authors are unaware of any past conference or journal literature on porting a chess engine to Android. We believe our contribution to the field to be novel in this regard, and this paper is meant to fill the gap. To readers interested in chess programming in general, we highly recommend the Chess Programming Wiki (2015), a rich online source of information on the subject.

[3] `https://play.google.com/store/apps/details?id=ccc.chess.gui.chessforall&hl=en`

[4] `https://play.google.com/store/apps/details?id=com.eng.engines&hl=en`

[5] Bik's Chess for Android, also compatible with his Chess Engines package, is also available on the Google Play Store and would also have been a good choice for this exercise. You can find it on the play store at `https://play.google.com/store/apps/details?id=com.google.android.chess&hl=en`. This choice was arbitrary, and should not be taken as any implication of one GUI being better than the other in any way.

[6] `https://play.google.com/store?hl=en`

2 Implementation

A few adjustments were needed to be made to the Firenzina code base in order to make the port work. These necessary adjustments are summarized below.

2.1 Build Changes

Android supports the same GNU tool chain already used to compile Firenzina for Linux, so most of the build files could be used without modification. The few minor adjustments needed were to select the correct tool chain to use for cross compilation and linking from a Windows on Intel host to an Android on ARM target. The makefile was initially configured to output a binary for Fire, upon which Firenzina is based, rather than Firenzina itself, so this was modified as well. Lastly, we changed the compiler options to specifically target Android, by telling it to tune for arm7 and removing the link to pthread, since it's already there by default (Google, 2014b), and the linker option fails otherwise.

2.2 Code Changes

The following code changes were required, almost all of which leveraged conditional compilation based on the __ANDROID__ define, which should always be there in the Android build system (Google, 2014c). In fire.h, a define was added to change Boolean to bool, and a "Plat" definition was added so that the engine reported itself as "Android" rather than "Linux 64". In win-linux.h, a define was added to change memalign to use an Android compatible memalign function, since it doesn't provide the POSIX compatible one. In utility.c, the NumCPUs variable was set by using Android's cpufeatures. Also, since cpufeatures (Google, 2014a) is normally provided via the static library compilation facility within NDK build, which we're not using, we copied over cpu-features.c and cpu-features.h and just compiled cpu-features.o as part of Firenzina itself, instead of building it as a separate static library.

3 Installation

To install Firenzina and play against it yourself, or to have it play against other engines, first install the Chess for All and Chess Engines programs from the Google Play Store. Next, copy the Firenzina executable over to your phone's SD card to a location you can navigate to easily later. Then, open up the "Chess Engines" program which should be located near the associated Chess for All. We use both of these to run our UCI chess engine via a GUI as described here, as shown in Figure 1.

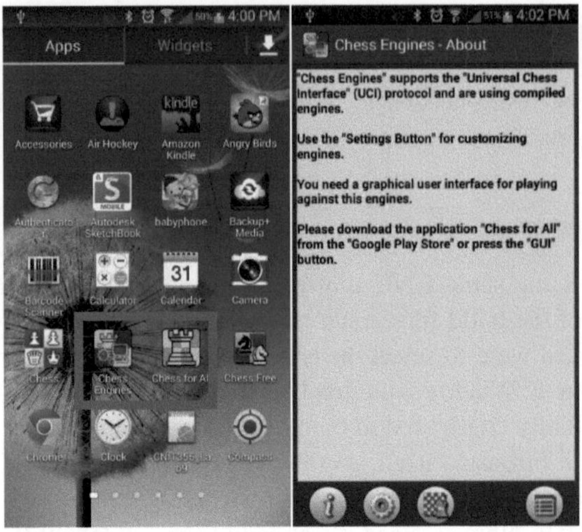

Fig. 1. Screenshots showing the Chess Engines and Chess for All icons on the Android apps list (left), and the Chess Engines – About screen (right)

From here, as shown in Figure 2, click on the small gear icon in the bottom menu bar, and then click "Install Engine from SD".

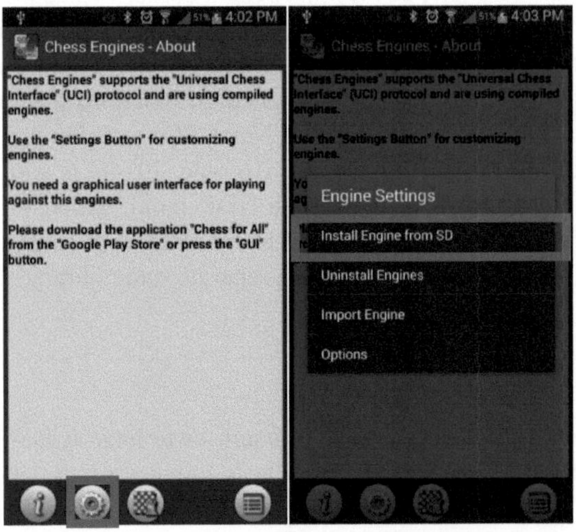

Fig. 2. Screenshots from the Chess Engines app, showing the icon to click on the Chess Engines – About page (left), and the "Install Engine from SD" button (right)

Choose the chess engine you want to use, which in this case is Firenzina as shown in Figure 3.

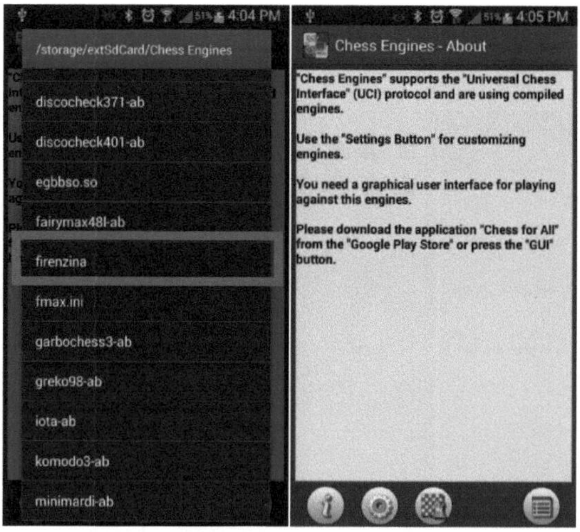

Fig. 3. Screenshots from the Chess Engines app, showing the engines list with **firenzina** highlighted (left), and the About screen, which is what you will see after choosing the engine

Now that you've selected it, you just have to import it as shown in Figure 4. Click the gear icon again and click "Import Engine". You should see the engine you have just installed in the list. In this particular case, it's Firenzina. Select it and click OK.

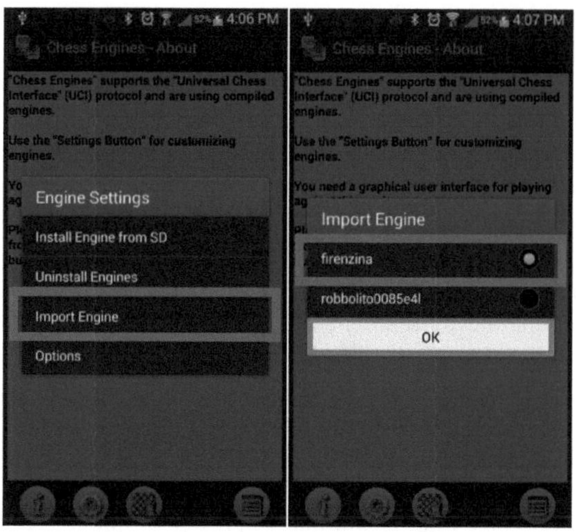

Fig. 4. Screenshots from the Chess Engines app, showing the "Import Engine" button (left), and the firenzina selection (right)

Now click the new game icon in the bottom tool bar, and it should launch Chess for All with this engine loaded as shown in Figure 5. You may have to configure Chess for All a bit to have it choose Chess Engines to play against and to run a new game, but as soon as that's done, you should be able to play against the engine, or even run a tournament of multiple engines on your Android smartphone or tablet.

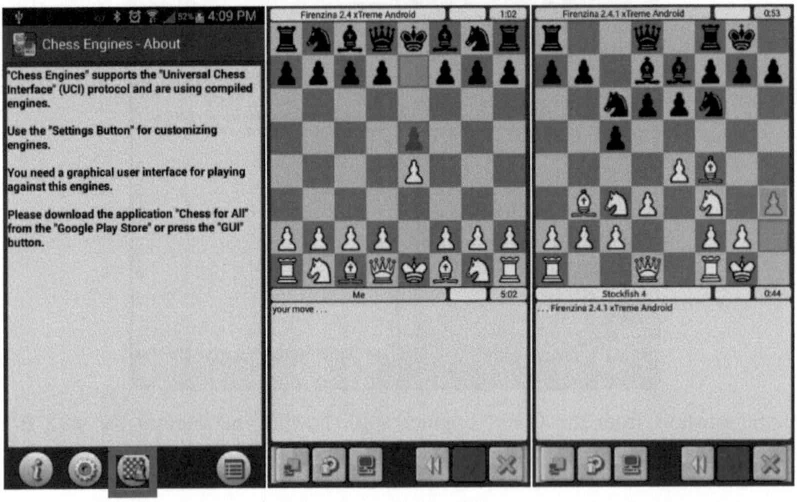

Fig. 5. A screenshot from Chess Engines showing the new game icon (left). Next we see a screenshot from Chess for All showing Firenzina playing against ourselves (center), and finally a screenshot of Firenzina playing against Stockfish 4 (right).

4 Performance

After the most recent prestigious Thoresen Chess Engine's Competition (TCEC), Season 7 (Thoresen, 2014), Fire, an older version of which Firenzina is based upon, ranked 7th behind other strong and well-known computer chess engines such as Stockfish, Komodo, and Houdini. Firenzina is not included in the TCEC tournament, so this is the closest we can get to assessing it there. (TCEC is often regarded as the unofficial world computer chess championship.)

However, in a separate tournament ran by Pohl (2014), the version of Firenzina upon which this Android port was based was included and placed 10th among similar competition. After filtering out closed-source engines from that list and considering the two versions of Stockfish as a single entity, Firenzina would be 3rd on this list among the open-source engines, behind only Stockfish and Gull. Gull would not be as easy to port to Android, as it relies upon processes instead of threads. This gives us some degree of confidence that the engine we ported, while not yet a tournament winner in its non-mobile version, is certainly competitive and a reasonable choice for our purposes. Also, Stockfish already has an Android port (Computer-Chess Wiki, 2015).

To gauge the strength of our Android port, we turn to Uckardes (2015a) RAPIDROID Android chess engine rating list. Here we see that Firenzina's rank has improved to 4th place, falling behind only Stockfish, Komodo, and Critter. While we recognize that several of the competitors from the non-mobile competitions do not seem to have ports included in this list, it is still encouraging that Firenzina places 4th out of 56 other Android chess engines, especially given that not much effort has been applied yet to tuning it specifically to this platform. By again filtering out the non-open-source engines from this list, we see that Firenzina places 2nd, only behind Stockfish. In a March 2015 update by Uckardes (2015b), we see that Firenzina has maintained its 4th place position, this time out of 104 total Android chess engines.

5 Software Engineering Analysis

While the above implementation and installation details make the port to Android seem relatively trivial, our experience in the process of getting to that point was actually somewhat painful. For instance, it wasn't always obvious why things like memory alignment issues were causing failures, or why the build was not completing. Getting it to the stage where it would execute at all from an Android command line, or even how to execute it on the Android command line, was not at all straightforward. Once we at least had it executing, attempts to track down errors with GDB were often not so helpful, but reverting to just adding simple printf statements was. Through the course of the port project, we added many lines of code while trying to troubleshoot and heading down false paths in our diagnosis of the issue, only to remove almost all of them later upon realizing at the end that there were in fact relatively few changes required at all.

The GUI tools we used were similarly nice once we had them working, but they also were not without their nuances to work through. For instance, it took some number of attempts before we realized basic mistakes we were making in trying to import the engine into the Chess Engines program to have it work with the Chess for All program. Similarly, when the program would crash arbitrarily while executing within that environment, debugging had to revert completely back to the tools mentioned above. So while in retrospect the installation instructions seem relatively simple, getting them to that point was not quite so easy.

Having said that, we feel this experience certainly beats at least one possible alternative, which would have been to try to re-develop Firenzina from scratch using the Java language directly as provided by Android, complete with our own custom user interface. While the NDK port took us several weeks to complete, we are confident that a complete rewrite would take months, if not years. Thus, the time invested in learning how to use the tools was well spent. Once we had the change set optimized, we found that, in order to implement this port, we only required modification of 12 lines and addition of 14 lines to the codebase of approximately 35k total lines.

Also, it is worth noting the value of the separation of business logic from presentation layer done so cleanly by modern chess engine protocols such as UCI. By adopting this architecture, GUI tools were able to be provided as independent apps, with which we were able to interface and have our engine play against both ourselves and other engines, which were also separate apps created by other developers. It was refreshing to us to see this technique's value so explicitly reinforced during this particular port. Other domains should consider ways to develop similar protocols to enable this level of separation, in order to reap similar rewards.

6 Conclusions

The main challenge with this port was in debugging. We were never able to get the debugger running to a satisfactory level during the port, but once we had resorted to printf style debugging, things went much better. Once we were able to narrow down the cause of the failures we were encountering and create the list of changes required, the port ended up being fairly straightforward. After removing all the extra code, we were also impressed by how relatively little needed to be changed to make Firenzina work on an Android smartphone. That speaks to both the cross-platform nature of the program, and the flexibility of the Android platform.

Chess engines in general, such as those based on standard protocols like UCI mentioned here, enjoy substantial benefit from separation of concerns in splitting business logic from presentation layer. Developers of new and existing application should consider implementing a similar protocol layer where possible, since as shown here, it enables rapid and modularized delivery of ported functionality.

Finally, our experience as described here in getting a chess engine running on Android reaffirms our belief in the usefulness of learning and making use of the porting tools available on platforms like Android. Especially when combined with sound architecture decisions such as the introduction of a protocol to help with separation of concerns, and the availability of other open-source components leveraging it, porting software to mobile is not only possible, in retrospect it may even seem easy. While the learning curve can seem steep at times, and the tools aren't always intuitive, at least for this data point, porting software to mobile rather than rewriting is the superior choice.

References

Bik, A.: Chessbase compatible engines. Aart's Blog (2013), `http://aartbik.blogspot.com/2013/07/chessbase-compatible-engines.html` (retrieved)

Chess Programming Wiki (2015), `https://chessprogramming.wikispaces.com/` (retrieved)

Computer-Chess Wiki (2015), Chess Engine List. Computer-Chess Wiki, `http://computer-chess.org/doku.php?id=computer_chess:wiki:lists:chess_engine_list` (retrieved)

Google, Android NDK CPU features detection library. Android NDK Documentation (2014a), `$NDK_HOME/docs/CPU-FEATURES.html` (retrieved)

Google, Bionic C library overview. Android NDK Documentation (2014b), `$NDK_HOME/docs/system/libc/OVERVIEW.html` (retrieved)

Google, Android NDK ChangeLog. Android NDK Documentation (2014c), $NDK_HOME/docs/CHANGES.html (retrieved)

Hyatt, R., Cozzie, A.: The Effect of Hash Signature Collisions in a Chess Program. ICGA Journal 28(3), 131–139 (2005)

Levy, D.N.L., Newborn, M.: How Computers Play Chess. W. H. Freeman & Co., New York (1990)

Liu, F.: Android native development kit cookbook. Packt Publishing, Birmingham (2013)

Meyer-Kahlen: UCI Protocol. Shredder Computer Chess (2000), http://www.shredderchess.com/chess-info/features/uci-universal-chess-interface.html (retrieved)

Pohl, S.: Bullet-ratinglist of the world's strongest engines. Stefan Pohl Computer Chess: private website for chess engine tests (2014), http://spcc.beepworld.de/top-bullet-list.htm (retrieved)

Ratabouil, S.: Android NDK: Beginner's guide. Packt Publishing, Birmingham (2012)

Schneirer, K.: Karl's Blog (2013), http://c4akarl.blogspot.com/ (retrieved)

Thoresen, M.: TCEC – Archive Mode. TCEC (2014), http://tcec.chessdom.com/archive.php (retrieved)

Uckardes, G.: RAPIDROID: Update 2015/01. Chesstroid: a chess experience for the Android user (2015a), http://chesstroid.blogspot.com/2015/01/rapidroid-update-201501.html (retrieved)

Uckardes, G.: RAPIDROID RELOADED: March-2015. Chesstroid: a chess experience for the Android user (2015b), http://chesstroid.blogspot.com/2015/03/rapidroid-reloaded-march-2015.html (retrieved)

SPEARS: Smart Phone Emergency and Accident Reporting System Using Social Network Service and Dijkstra's Algorithm on Android

Chakkrit Snae Namahoot[1,*] and Michael Brückner[2]

[1] Department of Computer Science and Information Technology, Faculty of Science, Naresuan University, Phitsanulok 65000, Thailand
chakkrits@nu.ac.th
[2] Department of Educational Technology and Communication, Faculty of Education, Naresuan University, Phitsanulok 65000n,Thailand
michaelb@nu.ac.th

Abstract. This paper presents research results on an efficient tool for emergency and accident reporting on Android smart phones with social network services. We have developed a location-aware Smart Phone Emergency and Accident Reporting System (SPEARS) for Facebook and Twitter users including agencies responsible for handling emergency situations and accidents. Agencies (e.g., police, fire departments, and hospitals) can store their locations via SPEARS, so that users involved in an emergency situation can retrieve the shortest path from the point of alert to the point of care, e.g. a hospital. On the other hand, agencies can retrieve the current location of the emergency via GPS and send help immediately. The shortest path is calculated by applying an improved Dijkstra's algorithm and displayed on Google maps in appropriate scaling. Moreover, SPEARS supports the collection and statistical presentation of emergency and accident data for further analysis.

Keywords: Emergency Response System, Android smart phone, Dijkstra's shortest path algorithm.

1 Introduction

In emergency situations, the traditional way is to call an emergency dispatcher (in Thailand 191 for general emergency calls, 1669 for medical emergency calls and 1155 for tourist related emergency calls). Besides trying to understand the emergency situation, the emergency dispatcher has to take notes of the location, which can be a problem because addresses can be confusing and not really well organized. A location-aware system harnessing an online map (e.g., Google Maps) can greatly improve mutual understanding of where the emergency happened or where to go next.

An emergency is an unanticipated situation, which may lead to the loss of life or property (gas explosions, collapsed buildings), to the harm of physical integrity of

* Corresponding author.

K.J. Kim and N. Wattanapongsakorn (eds.), *Mobile and Wireless Technology 2015*,
Lecture Notes in Electrical Engineering 310, DOI: 10.1007/978-3-662-47669-7_20

individuals or animals (car accidents), or to damage of property (fire, floods) or the environment (wildfire). Emergency situations are typically events that can be handled by local forces (police, ambulances, fire departments), whereas disasters are situations, in which damage with serious consequences occur on a regional or national level and affect the population as well as the emergency management forces themselves (cyclones, earthquakes). National or international assistance may then be needed to handle the disaster situation. A recent example of a disaster has been the 2004 tsunami, which affected several countries in Southeast and South Asia and led to a variety of emergency situations in the region that overloaded the emergency management forces. In this paper, we focus on emergency situations (for an appropriate test case, see Sect. 4.2).

An emergency plan is meant to prepare for a situation that cannot be controlled immediately through the cooperation of nearby and government agencies involved. There are three important reasons: 1) to help those who are in danger of life, build a treatment plan, and reduce the trauma following an emergency; 2) limit the damage to public and private property as well as the environment; and 3) to find the real cause of the emergency effectively.

Social Network Service is a form of website that builds a network for registered Internet users and is a communication channel at any time anywhere. This is why it is used for business advertising to promote organizations or companies, as well as a way to build relationships with customers to create events, or to request and provide information about products and services. It belongs to media of communication with organizations, which can be linked to internal communications and outside the organization effectively. Popular social networking sites include Facebook, Google Plus and Twitter. More and more of those services are also used for social improvement as such, e.g. improving peoples' lives or raising funds [1].

Google Maps is part of the Google search site which focuses on finding services, locations and routes. The Google Maps API is used to develop software to insert Google Maps into a component part of the user display, for example for emergency reports. Routes between relevant agencies and emergency places can then be shown.

When there is an emergency to report, much useful information on the emergency is not being released for the benefit of the public and for precautions in dealing with similar incidents. Considering the media to communicate and exchange information with each other in real time, social networks such Facebook and Twitter can be considered as very popular nowadays. Many organizations related to emergency events, including police stations began entering social network services like twitter and used to publish and report news to the public as well [2].

This research aims at developing an emergency reporting system with social network technology that can cause the user to update the data via Facebook or twitter via android smart phone. Google Map is used to display shortest routes of the relevant authorities to the occurred emergency in the quickest time .With Dijkstra's algorithm can be used to help finding the shortest path routing [3]. The developed system allows users to quickly report emergency situations that occurred in time and conveniently view any emergency. This will result in reducing damage to both body and property of the citizens of emergency happens. Also, information about or related emergencies reporting can be collected and stored in a database system for agencies or organizations to use in other areas of work or the various analysis. This can help to evaluate and plan to cope with emergencies and people life for more accurate and safety.

Dijkstra's algorithm is an algorithm used to find the shortest path. The algorithm is not too complicated and can be applied to find the shortest route on the map or the application in the network (see for example an implementation on [4]).

2 Literature Review

The recent advances in mobile communication and mobile information systems have had a significant impact on the development of Emergency Response Systems [5].

Yu et al. [6] have developed the Mapster application that uses Twitter to help in the event of emergencies in real-time .The application uses the retrieve details of the emergency from Twitter and display results with GPS via smart phones on the Windows Phone 7 platform. Moreover, the development also has the Open Data Kit to work with geography data in order to determine the extent of the interest in space and time of an emergency such as a flood can be viewed back on 10 hours ago within 30 km from the point where the user is enabling. Users can assess the situation and accurately predict events that might happen in the future to prepare for the events.

In the wake of the smart phone boom social network services were used at ever increasing levels. Soon, users began to use them as collaborative sources of information for partying and other social activities, including emergency reporting. Zach and Heverin [2] found Twitter playing a more significant role in the event of an emergency by exploiting real-time reports posted online. The research also showed that the use of Twitter can make contact of government officials with the public faster and leads to better public access of official information. Zach and Heverin used a survey of police stations in more than 30 major cities in the United States applying Twitter to exchange contact information with the public for about four weeks (they selected 4,915 tweets or 5,117 texts to be analyzed). They divided the texts into 11 categories, and the category with the most tweets consisted of crime and accident reports with 2,320 tweets, or nearly 45% of all tweets that were analyzed. This confirms that people report emergencies via Twitter as well. However, only 89 of the tweets did not fit into any of the categories.

Moreover, microblogging services are popular and use real-time Twitter messaging to disseminate information and build ad-hoc information networks. Okazaki et al. [7] studied the key features of a popular Microblogging service and conducted a project with a user-to-role-approach, imitating a sensor detecting an earthquake in Japan. There was a collection of all frequencies associated with the word "Earthquake" and "Shaking" in the tweets of users. Then a mathematical model with Kalman Filters was used to find the center of the earthquake from latitude and longitude of all tweet messages collected. The results were displayed on a map to warn warn the public of the earthquake faster and, effectively, in real-time. Some more examples are provided in Table 1.

Table 1. Characteristics of Emergency Response Systems

System	Characteristics	Reference
DUMBONET	Combination of satellite and wireless network with conventional Internet for disaster affected regions	Kanchanasut [8]
MIKoBOS	Provides mobile communication to fire departments for sharing information with colleagues and sensor data	Meissner [9]
MobileMap	GPS based mobile communication system between command center and firefighters for allocating human resources and materials	Monares [10]

The problem remains how rescuers and people in need can meet in the shortest time possible by using mobile technology. A well-known algorithm to discover the shortest path is Dijkstra's algorithm [11] or one of its many improvements. Panahi and Delavar [3] modified the shortest part algorithm using Dijkstra's algorithm for finding routes for emergency vehicles with a minimal time and real-time. The research aims to analyse problem of traffic congestion in the city, and the proposed route using GIS can be suggested and is able to solve network traffic congestions during driving. As a result, by providing dynamic shortest path and reducing the travel time and the passage of emergency vehicles is more efficient. Bauer et al. [12] have presented speed-up methods for Dijkstra's algorithm aiming at time scheduling on road networks in a number of different settings. For sparse graphs, they have found that a combination of contraction hierarchies [13] and arc-flags ([14], [15]) shows significant speed-up with low preprocessing resources.

3 System Design and Implementation

The system design of SPEARS has been focused on three main parts, which are explained in more detail in the following sections: the system architecture, the system implementation and the user Interface.

3.1 System Architecture

The components of the system architecture comprise (see Figure 1):

1. Users: log in via Facebook or Twitter for identification in order to access and use the system applications for an emergency reporting on Android (operating system) smart phones. The users can select Icon of various emergencies and can add more texts related. The users must be identified by phone numbers and names before reporting an emergency.
2. Web Server: to receive location information and messages in a form of JSON file from Twitter and Facebook and save them into the database and provide information in a form of Google Map to users and relevant agencies.
3. Database serves: to collect information about the emergency.
4. User Display: to show locations of the nearest relevant agencies where the nearest emergency incidents are reported on an Android smart phone.

5. Agency Display: to display locations where reported emergencies are received.
6. Administrators of relevant agencies can add their own locations and types of emergency and edit and delete locations whenever agencies have moved. The shortest routes can be printed out and handed over to officials who have to deal with emergency events.

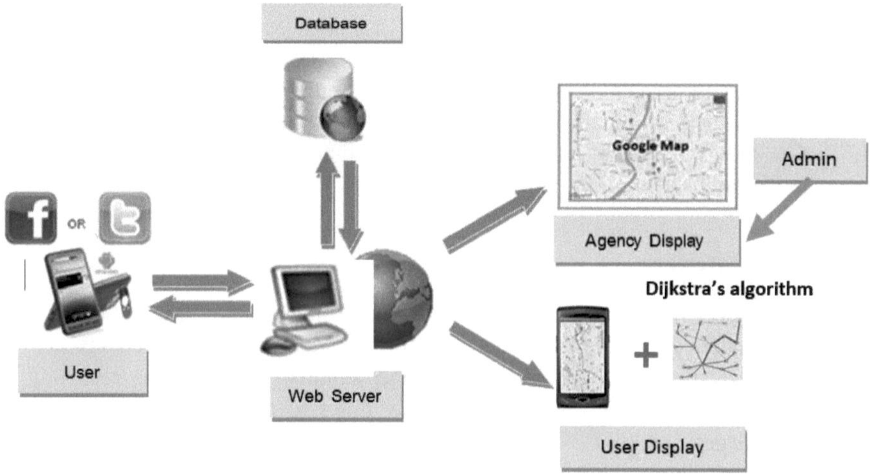

Fig. 1. System architecture

3.2 System Implementation

SPEARS has been developed using PHP and MySQL for web development applications. The Twitter API and the Facebook API can be used to retrieve data from Twitter and Facebook in JSON format and stored into database systems. Data can be received and sent between Browser and Server using AJAX. The Google Map API is used to insert a map on the Web Server, which can be displayed to users and relevant agencies via the Android operating system, which helps to report an emergency.

SPEARS displays this map of the emergency area in symbolic form and with colors; for example, a map relating a fire emergency uses the symbol [fire], and another relating a car accident uses the symbol [car crash]. The system can display the details of the reported emergency as well.

The system can calculate the shortest path from the relevant agencies to the point where emergency incidents are reported using modified Dijkstra's algorithm. Therefore, injured people can be taken to hospital in time, which can be a lifesaving endeavour. However, the system is limited to just the information within 30 kilometers of the point where the user is located and has reported an incident. Moreover, SPEARS can collect data from reporting emergencies that occur which can be analysed and shown as statistics.

The modified Dijkstra's algorithm can be described as follows:

Let the node at relevant agency's location be the initial node and let the emergency reporting point be the target node. First, the system checks the nearest relevant agencies and extracts all landmark nodes, distances and paths between the initial node and the target node from a database. The resulting data can be used in the Dijkstra's algorithm which calculates the shortest path as follows:

1. Set initial node to zero (current node) and all other nodes to infinity (unvisited nodes)
2. For the current node, find all unvisited neighbors and calculate their tentative distances. Compare the newly calculated tentative distance to the current assigned value and assign the lowest one.
3. Insert the lowest node into the visited path and set it as the current node.
4. Select only unvisited nodes and go back to step 2 until the target node has been visited, then terminate algorithm.
5. Get the shortest path from the initial to the target nodes with the smallest tentative distance of target node.

3.3 User Interface

Users can report incidents and emergencies via Facebook and Twitter but the users must be members of Twitter or Facebook before they can report emergencies.

Users must specify the phone number and name in the message reporting the emergency for identification. Users can view maps to relevant agencies and other reported emergencies.

1. Log in with Facebook or Twitter account
2. Inform types of emergency with names and phone numbers, e.g. fire, accidence, traffic jam without traffic light etc. or type in message explaining emergency if it is not shown.
3. Send an emergency location to relevant agencies and show the google map of the shortest route (using Dijkstra's algorithm) from the point where a user informs to agency locations and vice versa. These can help relevant agencies to get to the emergency location in time if there is any serious accidence. Also users can take injured people to the nearest hospital which they can view from the map via mobile. For example, if there is a fire reporting then relevant agencies such as fire station, police station, ambulance and rescue units will see the shortest route of fire location and can quickly deal with their own tasks efficiently.
4. Store emergency reports into web server for further analysis.

4 System Testing

We have tested the system using the city center of Phitsanulok province, Thailand, to be used to report an emergency. The data consists of two main parts: emergency situation and relevant agencies.

- Emergencies, such as traffic accidents, fire, electric power failures, and
- Relevant agencies, such as police stations, hospitals, rescue units, fire stations and electricity authorities. The system stores location data using longitude and latitude connected with Google Map. Examples of relevant agencies within Phitsanulok city are shown in Table 2 and build the basis of a first system test.

4.1 Mobile Interface

Figure 2 shows the mobile Interface of the emergency reporting system which allows users to log in using Facebook or Twitter (left picture) and to report the type of emergency (right picture). Those types of emergency buttons appear on mobile screen automatically according to the most frequent emergency situations in the region of report. As can be seen from top left clockwise, there are "Car Crash", "Traffic", "Violence" and "Crime". Apart from these emergencies, users are allowed to type in a message in the "Other" box. However, users must tick either the box "Injured" or "No injuries". Emergency buttons can be added and changed or modified by administrators as deemed useful.

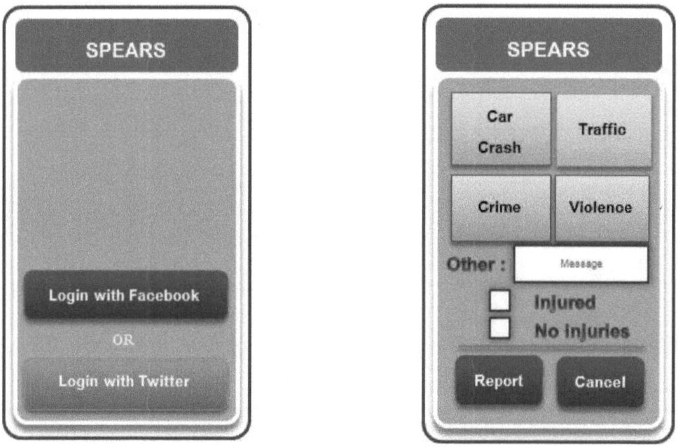

Fig. 2. The login Interface and types of emergency reports

4.2 A Use Scenario

An old woman has fallen over and is lying unconsciously on the temple premises of Wat Phra Si Rattana Mahathat Woramahawihan in the center of Phitsanulok, northern Thailand. A kind man is reporting by message "a faint woman" to SPEARS. The only hospital that appears on a map with the shortest route is "Ratanavej Hospital" (Figure 3, left picture), since the woman needs a hospital and the distance to this hospital is the shortest either to accompany by car or to be picked up by a hospital ambulance. Even though the "Fort Somdet Phra Naresuan Maharat Hospital" is the nearest beeline location, it results in a longer distance (calculated by Dijkstra's algorithm) and time since there is a one way route and people have to make a far U-turn. In addition, information and location of relevant agencies can be viewed (Fig 3, right picture) via SPEARS by pressing the "Agency Information" button.

After Rattanavej Hospital has got the message and the location, the system displays the shortest route and description where to pick up the fainting woman at Wat Phra Si Rattana Mahathat Woramahawihan (Fig. 4). Fig 4 shows the shortest route from accident location (Wat Phra Si Rattana Mahathat Woramahawihan) to Rattanavej Hospital or vice versa with additional information.

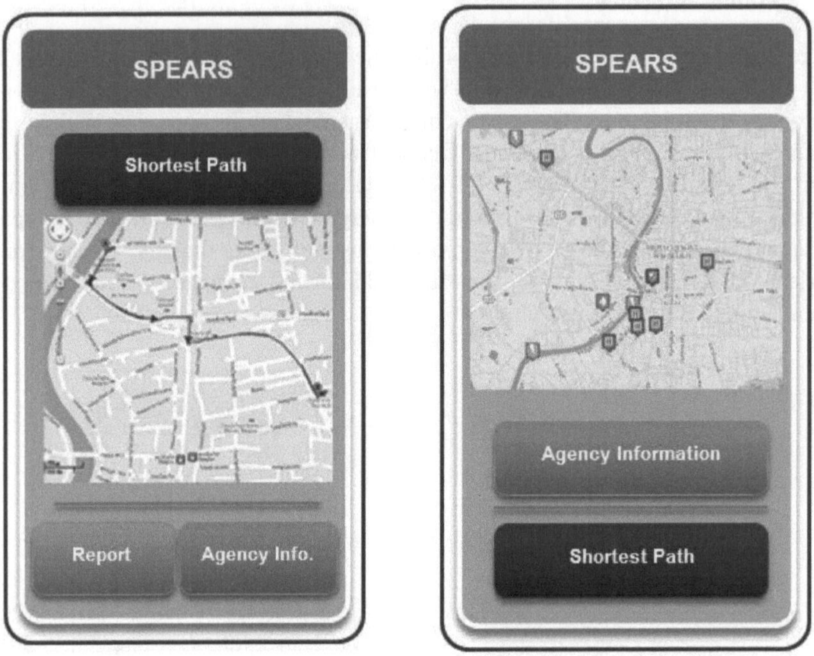

Fig. 3. Information on the shortest route and nearest agency via SPEARS

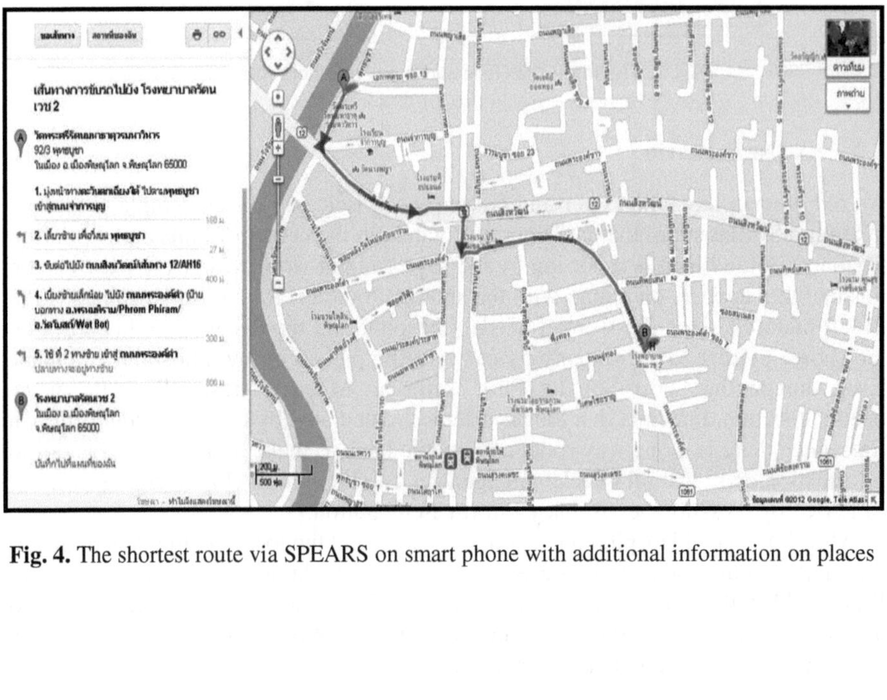

Fig. 4. The shortest route via SPEARS on smart phone with additional information on places

5 Conclusions and Further Work

In this paper we have reported on the results of a create-and-design research aimed at providing an Android based Smart Phone Emergency and Accident Reporting System (SPEARS) for Thai citizens and agencies responsible for rescue and emergency management.

In future versions of SPEARS, we plan to improve the system by letting responsible agencies decide to forward related information to other relevant agencies in case of emergency. Also, SPEARS will be able to deliver map based reports with emergency statistics showing hot-spots of accidents and emergency situations, which can support mitigation of such events.

A current limitation of the system is the use of Thai language only, which does not help much for most foreign tourists in the country. Moreover, SPEARS should enable users to communicate via streaming video to help emergency dispatchers understand the situation faster and in more detail.

References

1. Bresciani, S., Schmeil, A.: Social media platforms for social good. In: Proceedings of the 2012 6th IEEE International Conference on Digital Ecosystems and Technologies (DEST), Campione d'Italia, Italy, June 18-20 (2012)
2. Zach, L., Heverin, T.: Twitter for City Department Information Sharing (2010), http://www.asis.org/asist2010/proceedings/proceedings/ASIST_AM10/submissions/277_Final_Submission.pdf
3. Panahi, S., Delavar, M.R.: A GIS-based Dynamic Shortest Path Determination in Emergency Vehicles (2008), http://www.idosi.org/wasj/wasj3_(supplement%201)/14.pdf
4. Algolist (2010), http://www.algolist.com/code/java/Dijkstra's_algorithm
5. Machado, M.F.T., Nascimento, B.S., Vivacqua, A.S., Borges, M.R.S.: JEMF: A Framework for the Development of Mobile Systems for Emergency Management. In: Baloian, N., Burstein, F., Ogata, H., Santoro, F., Zurita, G. (eds.) CRIWG 2014. LNCS, vol. 8658, pp. 239–254. Springer, Heidelberg (2014)
6. Yu, L., Piyawongwisal, P., Handa, S., Liu, Y.: Going Beyond Citizen Data Collection with Mapster: A Mobile+Cloud Real-Time Citizen Science Experiment (2011), http://ieeexplore.ieee.org/xpl/login.jsp?tp=&arnumber=6130723&urlhttp%3A%2F%2Fieeexplore.ieee.org%2Fxpls%2Fabs_all.jsp%3Farnumber%3D6130723
7. Okazaki, M., Sakaki, T., Matsuo, Y.: Earthquake Shakes Twitter User: Real-time Event Detection by Social Sensors (2010), http://ymatsuo.com/papers/www2010.pdf
8. Kanchanasut, K., Tunpan, A., Abdul Awal, M., Kumar Das, D., Wongsaardsakul, T., Tsuchimoto, Y.: DUMBONET: a multimedia communication system for collaborative emergency response operations in disaster-affected areas. International Journal of Emergency Management 4, 670–681 (2007)

9. Meissner, A., Wang, Z., Putz, W., Grimmer, J.: MIKoBOS: A mobile information and communication system for emergency response. In: Proc. 3rd International ISCRAM Conference, Newark, New Jersey (2006)
10. Monares, A., Ochoa, S.F., Pino, J.A., Herskovic, V., Neyem, A.: MobileMap: A Collaborative Application to Support Emergency Situations in Urban Areas. In: Proc. of 13th International Conference on Computer Supported Cooperative Work in Design (CSCWD 2009), pp. 565–570. IEEE Press, Los Alamitos (2009)
11. Dijkstra, E.W.: A note on two problems in connexion with graphs. Numerische Mathematik 1, 269–271 (1959)
12. Bauer, R., Delling, D., Sanders, P., Schieferdecker, D., Schultes, D., Wagner, D.: Combining Hierarchical and Goal-Directed Speed-Up Techniques for Dijkstra's Algorithm. In: McGeoch, C.C. (ed.) WEA 2008. LNCS, vol. 5038, pp. 303–318. Springer, Heidelberg (2008)
13. Geisberger, R., Sanders, P., Schultes, D., Delling, D.: Contraction hierarchies: Faster and simpler hierarchical routing in road networks. In: McGeoch, C.C. (ed.) WEA 2008. LNCS, vol. 5038, pp. 319–333. Springer, Heidelberg (2008)
14. Lauther, U.: An extremely fast, exact algorithm for finding shortest paths in static networks with geographical background. In: Geoinformation und Mobilität - von der Forschung zur praktischen Anwendung, vol. 22, pp. 219–230. IfGI (2004)
15. Möhring, R.H., Schilling, H., Schütz, B., Wagner, D., Willhalm, T.: Partitioning graphs to speed up Dijkstra's algorithm. In: Nikoletseas, S.E. (ed.) WEA 2005. LNCS, vol. 3503, pp. 189–202. Springer, Heidelberg (2005)

Descubre PUCP: Mobile App to Improve Academic Experience Inside Campus

Pedro Jesús Carrión Castagnola, Natalí Flores-Lafosse, and Albert Díaz-Mauricio

Dirección de Informática Académica, Pontificia Universidad Católica del Perú
Av. Universitaria 1801 San Miguel, Lima, Perú
{pjcarrion,natali.flores,adiaz}@pucp.edu.pe

Abstract. Descubre PUCP is a project developed and implemented in the Pontificia Universidad Católica del Perú whose objective is the improvement and extension of academic experience in college outside the classroom. To achieve this, Descubre PUCP marks each place in campus as a point of interest, showing its location using an augmented reality interface or a Google Map, adding relevant information for each one. This information includes static data (descriptions, photos, telephone numbers and mail contact) and dynamic data (in real time), interacting with different services and applications of PUCP such as events about to start in the auditoriums, free computers in the laboratories, the dishes to be served in the cafeteria, the streaming transmission available in different congresses or seminars in the campus and also explaining how to get to these places.

Keywords: augmented reality, point of interest, real time information, maps, streaming, minimal route.

1 Introduction

When a student or staff member just becomes part of a college community, it's understandable that he doesn't know where many places of campus are. These include some academic buildings, staff offices, libraries, cafeteria or general services. Even people who are already familiar with the campus, may don't know places that have recently been inaugurated or haven't visited before.

It's also common for a college to be the site of multiple events both for the members of the university and third parties. But it's usual that all the events don't have enough visibility among the community even when a great part of them could want to be present. It's also a common problem, mostly along the students, that when they need a computer they don't know where one is available in the laboratories.

All these situations require a solution that allows the student and other members of the college community to connect with campus information in an easy and intuitive way. As Diana G. Obliger [09] mentions, the goal is to "connect points to form roads", in reference of the information that we receive and later get integrated in our everyday lives. This integration is what a mobile device uses to solve these limitations.

K.J. Kim and N. Wattanapongsakorn (eds.), *Mobile and Wireless Technology 2015*,
Lecture Notes in Electrical Engineering 310, DOI: 10.1007/978-3-662-47669-7_21

It's worth adding that the mobile devices and applications' use are already part of the ecosystem of a modern campus, in which services get a context according a determined situation. That is, an application may be present in each moment of a user and as needed, bring him what he wants to know. For example, is he hungry? Know where the nearest cafeteria is located and what dishes are they serving. Looking for a PC to print some paper? Find there is a free computer in a determined lab. Is he near an auditorium? Discover there is a conference near with a streaming that may be of interest.

2 Previous Works on Augmented Reality

Augmented Reality (AR), that is, the supplementing of information in a direct view, has been a field of research since its first appearance in works of George Stratton [12][13], where only some change to the vision was made, going up to more recent technology developments in charge of Steve Mann [07][08], focused in wearables and helping people with some visual impairment. There is also some spotlight in devices such as Google Glass and similar [10], even when some of the support has decreased in last years. The wide range of applications of AR brings a lot of possibilities of use to help people unfamiliar to some places and even bring more useful information of the surroundings.

There have been some projects surrounding this same challenge but in other environments. For example, Hui, Lin et al [05] details a mobile app that helps tourist get information of the different Points of Interest (POI) surrounding the sites of the Yilan Hot Springs. They use capabilities provided by Google Maps and OmniGuider GPS Navigation System to help tourist self-guide themselves when they arrive at location.

2.1 Mobile Applications for Universities

With the ever-growing mobile application offer and application downloads [16], it is important for organizations to create a visual presence for these interfaces, either by a responsive web or some kind of mobile app. Universities can't escape this reality, so naturally more mobile applications appear to aid students, faculty and staff to obtain in real time different information.

Some efforts have been made around the globe. Delail et al [02] presented a case study of an indoor AR mobile application that gives the user the ability to share location, find places and voice-command search. It was developed for iPhone only. Another Campus Mobile App, for Android, was developed by Wong, L.H. [15] using partly AR but also including learning modules to complement the experience and give a more educational approach. It's interesting to notice the feedback analysis that was made to evaluate the usability of the application.

In the context of the Pontificia Universidad Católica del Perú, another mobile application was distributed at a later date than Descubre PUCP was made available. This application focused more on the courses (times, grades and documents) where students are enrolled in, and complements the everyday information that Descubre PUCP brings for a more administrative approach,

3 Descubre PUCP Features

Descubre PUCP is a mobile application that brings relevant information to college community using modern technologies, with the capacity to be updated and get information in real time. It's developed in Android, taking in account its participation in the tablet's market[1] and the more accessible prices for students. It is important to note that the openness policy of the O.S allows to more flexibility in the development of the application.

Among the principal features of Descubre PUCP are the geo-localization tools and the real time information gathering. These will be explained in detail.

3.1 Geo-localization Tools

Augmented reality is a term that defines technology capable of adding virtual elements in the visualization of an environment by and interface, allowing the user to feel that virtual and real objects co-exist in the same place. This project uses this technology to show information of the place that has the current focus [01].

Google Maps is a service provided by Google that allows developers to use an API in their own applications. It allows navigating over maps and provides detailed information about routes, satellite imaging and important places names, among others. These maps are used by Descubre PUCP to show the location of points of interest in a map [03].

Fig. 1. List and map of the university showing a point of interest (a) and a point of interest with route (b) created with the Google Maps for Android libraries

[1] The cumulative downloads of applications since half of 2010 to half of 2013 in the most popular app stores: Google Play y App Store, go to 48 and 50 billion each [14].

Mixare [06] is a free open source library that allows third party applications to use augmented reality in their own applications. It integrates with the geodata source of the developer. Descubre PUCP uses this library to display the augmented reality window inside the campus.

Fig. 2. Augmented reality viewer of Descubre PUCP, inside the campus

3.2 Point of Interest's Information from the Server

By communicating with a server, Descubre PUCP shows the following information of each point of interest: name, description, service hours, links, parent place, related places, contact (phone, e-mail address or more links), photos and location references. The places, phone numbers and e-mail addresses are selectable and allow showing information of the place, making a call or sending a default message by the smartphone, respectively. All this information is added by the system administrator using a web interface.

Fig. 3. An information example of a point of interest in Descubre PUCP

The information shown depends of the place that is accessed. Some places require some real-time information to be displayed. Some of these are the computer laboratories (it shows the number of available PC), cafeteria (it shows the dishes schedule) and auditoriums (it shows the events information and event streaming if available).

Fig. 4. Window showing the availability of computers in each laboratory (a) and the programmed meals in a cafeteria (b)

The application also allows the activation of notification of "near events". That is, when the user walks near a place where an event if about to start, a notification of the event will be triggered to inform him.

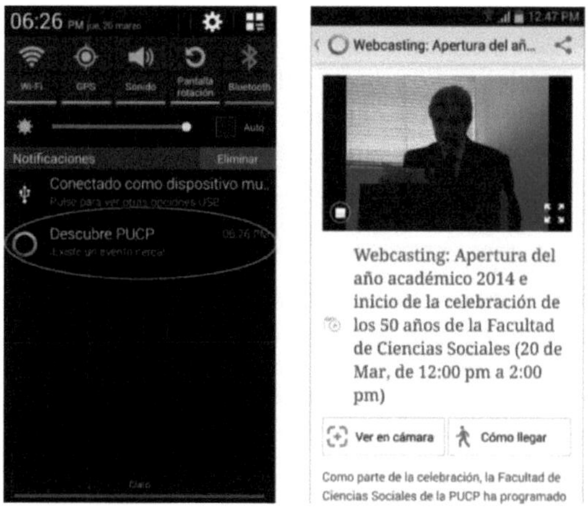

Fig. 5. Window showing the "near event" notification (a) and the detail of a streaming option (b)

Given the ease of use and net velocity, all the information provided by Descubre PUCP is obtained using a connection by REST [04] web services. The file format for that is JSON.

4 Design and Implementation

For this project development, it was decided to construct a native mobile application for Android that uses information of its own server.

4.1 Architecture

The solution's architecture defines the higher level design of the project. A graphic that shows this architecture follows, with an explanation of each component below.

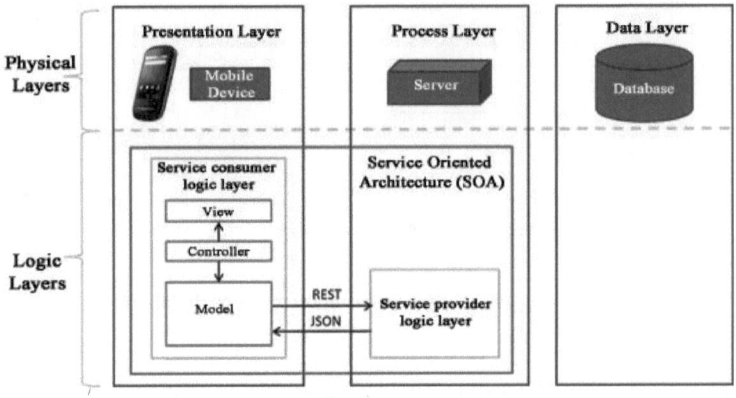

Fig. 6. Architecture diagram of Descubre PUCP

Physical layers: A 3-layer hardware architecture[2] was selected: presentation layer, process layer and data layer.

- Presentation Layer: Defined by the mobile app views.
- Process Layer: Conformed by the data server where the necessary tools are installed to get information from the database, construct and organize it in JSON format and display it. This layer is where the images of points of interest are stored and provides the interface of the system administrator to input the information of the point of interest.
- Data layer: Conformed by a database that contains all the information of the points of interest. This information is delivered to the Process layer every time it's called.

Logical layers: For the software architecture, it was decided to use a Service oriented Architecture[3], meaning web services will be implemented.

The best advantage of this type of architecture is that allows the communication of different application independently of how they were built. With this, we may provide information to a mobile application, a web site or other applications as needed.

This architecture allows identifying two logic layers: service provider logic layer and service consumer logic layer.

- Service provider logic layer: Conformed by a php-written program which is responsible to validate the parameters sent by HTTP (as we established a REST web service will be used), build and execute the necessary queries to the database and print the information of the point of interest in JSON format.

[2] Also know ad 3 Tier Application.

[3] Also known as SOA.

As shown in the Architecture diagram, this flow occurs by communicating with and receiving from the model defined in the service consumer logic layer.

Additionally, this layer is responsible of providing the administrator an interface to create, edit and delete points of interest.

For the real-time information, this layer is the one that processes and stores automatically the information provided by the Web Services. Time between each information recollection depends on the information type. For example, the information of free PCs in the laboratories is of 45 minutes, while the events information is collected once a day.

– Service consumer logic layer: Conformed by the mobile app developed for Android OS. Its structure was made following a Model, View, Controller pattern (MVC). This pattern' main characteristic is to give independence to the application's logic from the interface design and information structure. These layers communicate with each other using the controller as an intermediary.

- Model: Represents the connection with the server information
- View: Represents all the needed resources to show the user interface. Using the Android SDK, the design and architecture are implemented using XML structures.
- Controller: This layer is represented by the application logic, meaning the classes that implement the processes and calculation necessary to accomplish the features of the application. It responds to the user's actions and is in charge of calling the requests to the Model and View.

4.2 Information Architecture

This section analyzes and gives the foundations used for the organization, disposition and presentation of data.

Navigation: The application has 3 levels of depth, represented by the start window, points of interest window and information of a point of interest. The augmented reality and map windows are considered extra-information windows and may be accessed from any level. The help window may only be accessed from the start window.

The following diagram shows the navigation between windows in the application. Arrows show which windows may be opened from each window and the action needed for doing it.

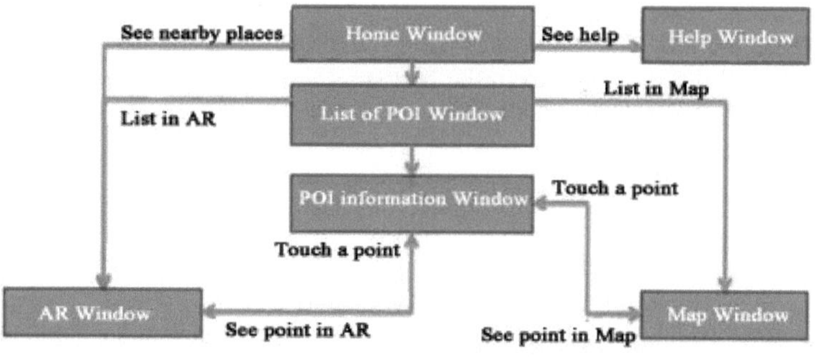

Fig. 7. Navigation diagram of applications' windows

5 Distribution and Feedback

The application was made available in Google Play to every public, being the only requisite for installation that the device had a camera (for the augmented reality to work). The first version only included the maps and AR features, being followed by a version that included the dishes schedule.

The next update incorporated the events, notifications and streaming features, as an API was made available by the event-streaming services of the university. The last update made incorporated the real-time visualization of the free computers in the laboratories.

In the last year, some efforts have been made to measure the impact of the application in the university community in the form of random interviews and a focus group. The results have given favorable reviews but a lack of knowledge of the existence of the application by much of the community. In general, they found the application very useful but weren't aware of its availability in most cases.

6 Conclusions and Future Works

Even when this project hasn't a properly academic approach, it indirectly intervenes in the lives of students and staff as it brings information to the university community that saves time in finding places and information, which improves the academic experience in campus.

About two years ago, Descubre PUCP was made available by the Pontificia Universidad Católica del Perú, and has reached more of seven thousand downloads in the application store (Google Play). There is currently work being done to develop a version for IOs devices, and a constant research of new features and types of information that could be useful for the university community. The usage of the data that the application receives is also a factor to be exploited to give a more personalized experience to the user. [11]

A very strong effort is also being made to make the application known around more people in the campus, specially targeting the first year students, who are by far the ones for which the application would be of most usefulness.

Acknowledgements. The authors wish to express their thanks to the engineer Jackeline Trujillo for her support in the first stages of development of the application. Also, thank Mr. Andres Ghersi for his collaboration in the development of the web interface of the application.

References

1. Azuma, R.: A Survey of Augmented Reality, California, pp. 1–3 (1997)
2. Delail, B.A., Weruaga, L., Jamal Zemerly, M.: CAViAR: Context aware visual indoor augmented reality for a university campus. In: Proceedings of the The 2012 IEEE/WIC/ACM International Joint Conferences on Web Intelligence and Intelligent Agent Technology, vol. 03. IEEE Computer Society (2012)

3. Google Maps Android API, http://developer.android.com/google/play-services/maps.html
4. Francia, S.: SPF13 "SOAP vs. REST" (2010), http://spf13.com/post/soap-vs-rest
5. Hui, L., et al.: Mobile Augmented Reality of Tourism-Yilan Hot Spring. In: 2014 7th International Conference on Ubi-Media Computing and Workshops (UMEDIA). IEEE (2014)
6. Mixare (2015), http://www.mixare.org/
7. Mann, S.: Mediated reality with implementations for everyday life. Presence Connect, August 6 (2002)
8. Mann, S.: Eye am a camera: Surveillance and sousveillance in the glassage. Time Magazine 2 (2012)
9. Oblinger, D.: Higher Education in the Connected Age. In: Diggs, T. (ed.) EDUCAUSE Review, pp. 4 a 6. Educause (September/October 2013)
10. Sangani, K.: Developing augmented reality apps (2013), http://eandt.theiet.org/magazine/2013/04/developing-ar-apps.cfm
11. Smith, A.: Location-based apps present opportunities – and data challenges (2014), http://searchcrm.techtarget.com/feature/Location-based-apps-present-opportunities-and-data-challenges
12. Stratton, G.: Some preliminary experiments on vision. Psychological Review (1896)
13. Stratton, G.: Vision without inversion of the retinal image. Psychological Review (1897)
14. Techcrunch. The App Store's 50B Downloads Vs. Google Play's 48B: Android Closes The Gap, http://techcrunch.com/2013/05/15/the-app-stores-50b-downloads-vs-google-plays-48b-android-closes-the-gap
15. Wong, L.H.: Mobile Campus Touring System based on AR and GPS: a Case Study of Campus Cultural Activity. In: Proceedings of the 21st International Conference on Computers in Education. Asia-Pacific Society for Computers in Education, Indonesia (2013)
16. Xataka, G.: Se vendieron 195 millones de tablets en 2013, el 62% eran Android, http://www.xataka.com/tablets/gartner-se-vendieron-195-millones-de-tablets-en-2013-62-eran-android

Localizing a Flying Object on Target Place Using Heterogeneous Binary Sensors

Hyunyoung Kim[1], Changhee Cho[2], Jisu Kim[1], Sanghyun Park[1], Jinsul Kim[1,*], and Kuinam J. Kim[3]

[1] School of Electronics & Computer Engineering, Chonnam National University, Gwangju, 500-757, Korea
[2] Graduate School of Interdisciplinary Program of E-Commerce, Chonnam National University, Gwangju, 500-757, Korea
[3] Department of Industrial Security, Kyonggi University, 94-6 Yiui-dong, Yeongtong-gu, Suwon, Kyonggi-do, 443-760, Republic of Korea
{hyuny.kim,jsworld}@chonnam.ac.kr, xisom@xisom.com, {sanghyun079,dyrk10}@gmail.com, harap123@daum.net

Abstract. In this paper, we propose a spatial and temporal flying target detection method using different types of sensors. By combining the sensors, we can discover the hit time and location more precisely. We designed ta system composition in consideration of protecting sensor system, and implemented a working prototype. In the experiments, we demonstrate its performance depending on the processing module and the accuracy.

Keywords: flying object detection, vibration sensor, infrared sensor, interpolation.

1 Introduction

Detecting a moving object from images is an important research area in modern military as well as in many other applications. When it comes to detect a flying projectile nearby the target area, sensor installation location becomes important. As a projectile can hit the sensor and dust can spatter to the sensor, the sensor should be placed as far from the target.

In this paper, we introduce a system for high-speed projectile detection. The sensor are elaborately placed to avoid possible destruction. By using heterogeneous sensors, we localize the projectile on target. Related system comparison and interpolation method are provided.

2 Related Work

Bae [1] suggested a small target detection technology using bilateral filter and temporal cross product. This technology is specialized in detecting a target in sky among cloud. Weng [2] and his colleagues introduced a weather-adaptive flying

* Corresponding author.

K.J. Kim and N. Wattanapongsakorn (eds.), *Mobile and Wireless Technology 2015*, Lecture Notes in Electrical Engineering 310, DOI: 10.1007/978-3-662-47669-7_22

target detection. They adopted fuzzy system to classify weather condition and the target status, and used a Kalman filter to predict the target location. Their system shows very high accuracy in clear weather and 90% accuracy in the other case.

Kim [3] After installation, distributed sensor in a space, obtained in a way that keeps track of the weight assigned to a higher value close to the target with a binary proximity sensor. Shrivastava [4] OCCAMTRACK to calculate the velocity estimate associated with the piecewise linear trajectory algorithm development and geometric particle filter algorithm and post-processing via the track a target with the least amount of information obtained from the binary sensor.

3 Target Detection with Vibration Sensors and an IR Array

A number of constrains were placed upon the overall sensor system. The primary of these were that it should be placed out of the possible object's flying path, since an object is highly possible to hit anything in its moving direction and destroy the system. We use a series of binary sensors and IR sensors to detect the object. The binary sensors are placed on backside of the target so that they detect the impact on the target. IR sensors are installed only on bottom side of the target. By using proper case and keeping distance between the sensors and the target, we can reduce the possibility for hit.

The secondary issue is the speed of the object. Even though we limit the object to a projectile, its speed varies on where it is used, what the firearm is, and what the projectile is. For example, the speed in a shooting Olympic is from 100 ~ 270 m/s. The speed of army bullet is far faster, over than 900 m/s when it is shot by K2. The sensing rate of system should be high enough to detect projectiles. Once it can detect the projectiles in army, it can detect others for game.

3.1 System Design

Figure 1 shows the installation example of heterogeneous sensors. The target size is fixed to 500 by 900 mm as used in the National Police Agency. Binary sensors are

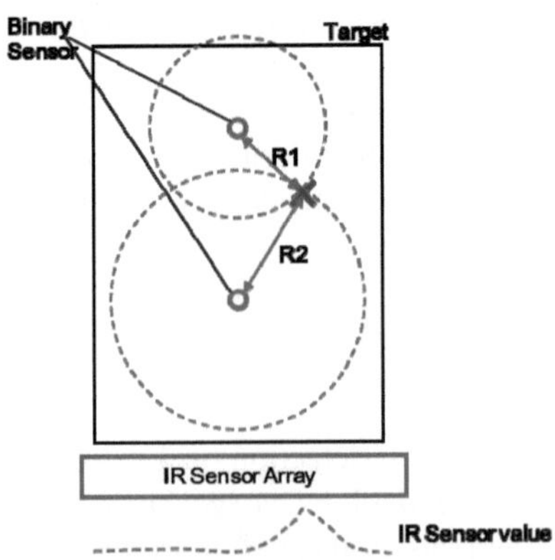

Fig. 1. Sensor installation design. Two vibration binary sensors and an array of infrared sensors are used.

installed backward and the IR sensors are placed under the target. They continuously read their status. When an object hits the target, the binary sensors detect the impact and can infer the hit location by the impact degree. When an object pass above the IR sensors, it reflects the IR light from the sources and the IR receiver can detect the current reflected light intensity.

3.2 Implementation

We implemented the system and the evaluation environment. Two gyro sensors were used as vibration sensor and installed on backside of the target. When projectile hit the target, the impact generates vibration on the target and propagated to the sensors. The vibration strength is related to the distance between the projectile and the sensor.

IR sensor array is installed below the target. The distance between the array and the target can be adjusted by using strong IR emitters. By giving some distance, there is lower possibility to destroy the array. A frame can be applied to array, perpendicular direction of the projectile so that the array can be shielded.

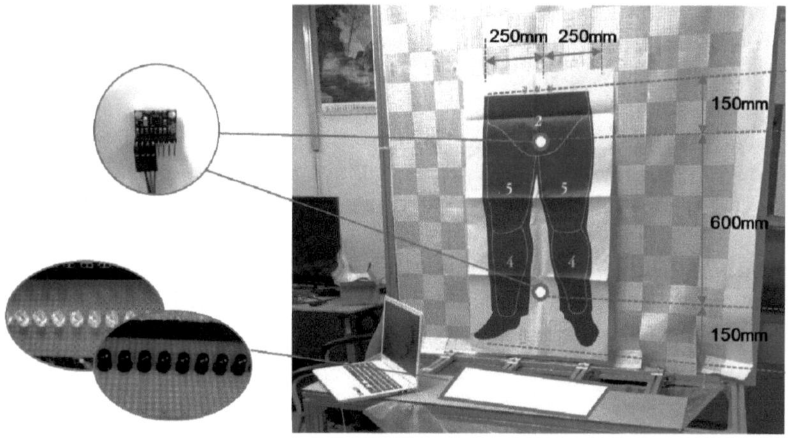

Fig. 2. Sensor system implementation. Vibration sensor are installed the backside of the target of the IR emitter and receiver array are installed below the target.

4 Evaluation

ADC Performance Evaluation. The analog-digital conversion speed significantly affect the performance of projection sensor system. To detect an object with 960m/s speed by each centimeter, the infrared sensor array should read the all sensor values every 10us. In our prototype, we set our goal to detect a BB by a popgun. We iterated the code 100 times for each condition. For internal ADC elements, 8 infrared sensors were read for once and 2 infrared sensor were used for external ADC sensor. We calculated the time for one analog-digital converting by (time for each ADC) = (Total time consumed) / ((number of iteration) * (number of sensors)). The results are shown in table 1.

Table 1. Performance comparison among analogy-digital converter (ADC) elements and division factors (DF)

ADC Type	ADC element	128 DF	16 DF
Internal ADC	Atmega 328 (Aruino Uno)	112us	17us
Internal ADC	Atmega 2560 (Arduino Mega)	112us	17us
Internal ADC	ARM Cortex-M3 (Arduino Due)	4us	-
External ADC + MUX/DEMUX	MCP3201+CD4067	67us	-

ADC converting time highly depends on the CPU clock rate when the CPU types were similar. The two different AD converter, Atmega 328 and Atmega 2560 showed the same result because their CPU clock were same to 16MHz. ARM Cortex-M3 showed the highest performance as it had 84MHz of CPU clock. However, it had 5.25 times higher CPU clock but the ADC converting was 28 times higher.

A lower division Factor (DF) can raise ADC speed. We accessed the Atmega's register and changed the division factor and got faster converting time. There is a tradeoff between the ADC speed and accuracy in this method.

We assumed that high speed will be enabled by allowing simultaneous AD converting by using external ADC. The converting time for one ADC was short but the channel alternating in multiplexer/ de-multiplexer caused extra time. This could be reduced by introducing a small status storage between an ADC and a multiplexer.

Interpolation. Interpolation methods determines the object position. We used spline interpolation method and acquired the image as demonstrated in Figure 3. We can acquired the position of the passing object.

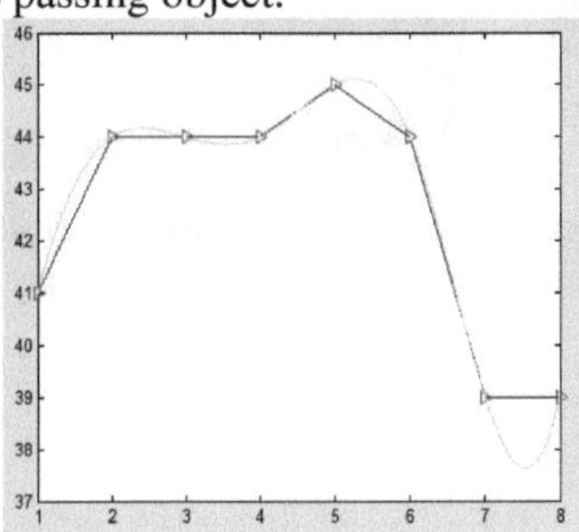

Fig. 3. An example of interpolation. Red triangles are sensor valued. Blue line shows a linear interpolation, and green line shows a spline interpolating.

5 Conclusions

This paper introduced a method to detect flying object using heterogeneous binary sensors: vibration and infrared sensors. The sensors are elaborately installed to avoid destruction by the object. Sensing methods for the system is demonstrated and evaluated. We found more suitable microcontroller for high speed projectile and a interpolation method to find projectile location.

Acknowledgments. This research was financially supported by Korea Evaluation Institute of Industrial Technology Research Program, Korea 2014.

References

1. Bae, T.-W.: Small target detection using bilateral filter and temporal cross product in infrared images. Infrared Physics & Technology 54, 403–411 (2011)
2. Weng, T.-L., Wang, Y.-Y., Ho, Z.-Y., Sun, Y.-N.: Weather-adaptive flying target detection and tracking from infrared video sequences. Expert Systems with Applications 37(2009), 1666–1675 (2010)
3. Kim, W., Mechitov, K., Choi, J.Y., Ham, S.: On target tracking with binary proximity sensors. In: Fourth International Symposium on Information Processing in Sensor Networks, pp. 301–308. IEEE (2005)
4. Shrivastava, N., Madhow, R.M.U., Suri, S.: Target tracking with binary proximity sensors: fundamental limits, minimal descriptions, and algorithms. In: Proceedings of the 4th International Conference on Embedded Networked Sensor Systems, pp. 251–264 (2006)
5. Perrin, F., et al.: Mapping of scalp potentials by surface spline interpolation. Electroencephalography and Clinical Neurophysiology, 75–81 (1987)
6. Shan, Y., Speich, J.E., Leang, K.K.: Low-cost IR reflective sensors for submicrolevel position measurement and control. IEEE/ASME Transactions on Mechatronics 13(6), 700–709 (2008)
7. Zizka, J., Olwal, A., Raskar, R.: SpeckleSense: Fast, Precise, Low-cost and Compact Motion Sensing Using Laser Speckle. In: Proceedings of the 24th Annual ACM Symposium on User Interface Software and Technology, pp. 489–498 (2011)
8. Eisenhardt, P.R., Stern, D., Brodwin, M., Fazio, G.G., Rieke, G.H., Rieke, M.J., ... Tiede, G.P.: The Infrared Array Camera (IRAC) Shallow Survey. The Astrophysical Journal Supplement Series 154(1), 48 (2004)

Biometric Authentication Technology Trends in Smart Device Environment

HyunJin Kim[1], JunHoo Park[1], JangYong Lee[2], and JaeCheol Ryou[1]

[1] Department of Computer Engineering, Chungnam National University, Daejeon, Korea
[2] The 3rd R&D Institute-4, Agency for Defense Developments, Korea
be.successor@gmail.com, {jhpark,jcryou}@home.cnu.ac.kr, flukelee@add.re.kr

Abstract. In a computing environment, Authentication is one of the most important techniques. ID/Password scheme which is currently the most widely used authentication methods is easy to implement and the scheme have the advantage of low cost, etc., but there are many issues from a safety. OTP(One Time Password) is characterized in that a disadvantage in usability, on the other hand, with a number of advantages for security. Among these recent years, Biometric authentication is attracting attention as the spread of smart device equipped with a biometric smart sensor. In this article, we will analyze the limit of knowledge-based authentication and possession-based authentication and the prospect of a biometric authentication technology. And we will analyze the FIDO protocol that provides a user authentication function using a biometric technique that is used by the smart devices in the online environment.

1 Introduction

Authentication is a technology that computer user in contact with most often. In the off-line, it is possible to determine the other by face to face, but in the non-face-face environment, it must be mobilized the various means to verify the other party. And this cryptography technology is accompanied by essential to them.

Authentication components to verify the users are classified into "Knowledge-based authentication" largely based on the memory of the user, "Possession-based authentication" based on the user's possession medium, "Bio-based authentication" based on the user's Biometric information. Among those authentication components, "ID/Password authentication" of one of the "Knowledge-based authentication" scheme is currently the most commonly used authentication means that is easy to implement and have the advantages of low cost, etc. [1].

ID/Password is performed by verifying whether a password matches the ID that identifies the user. Therefore, password is a secret information and it is a principle that password must be generated unknown status. For this reason, In a number of systems are recommended to use a mixture of 8 digits and more various types of letters and characters(number, capital letter, small letter, symbol, etc.). However, most people uses password as a combination of the date of birth, anniversaries, and phone numbers because these are easy to remember. Also they use the same ID/

K.J. Kim and N. Wattanapongsakorn (eds.), *Mobile and Wireless Technology 2015*,
Lecture Notes in Electrical Engineering 310, DOI: 10.1007/978-3-662-47669-7_23

Password for many systems, as well as repeatedly. So as, many social engineering problems are accompanied by ID/Password [2].

Thus, in ID/Password methods, the user's role is important as using a complex passwords, using the different passwords, periodically changing passwords. In addition to, aspect of system, user should be set the validity and periodically the validity change through the management policies such as a Password Aging.[3] But ID/Password are vulnerable to cyber security threats, such as Brute force attack (Brute-force Attack), a dictionary attack(Dictionary Attack), backdoors, etc. Although the technical and managerial responses such as restricting the number of login, vaccines, etc. exist, thereby it is happened that benefits such as facilitating implementation, low cost are reduced [4].

In chapter 2 of this paper/article, we will examine the status of electronic authentication technologies in the financial sector, which is one of the most important. In section 3, we will look at the view of biometric authentication technology that is attracting attention as a next-generation alternative to knowledge-based authentication and ownership-based authentication. In section 4, we will analyze the FIDO protocol, one of the protocols, using biometric technology in environment of smart devices. And in section 5, we will draw a conclusion.

2 Electronic Banking Authentication Technology

Financial sector is fields where authentication is one of the most important in. In Korea, financial companies are induced to use the "OTP" as a means to enhance the security and to reduce the fraud accidents. Therefore, in Korea, as shown in figure 1, the number of Korea's OTP deal and the number of Korea's issuance of OTP has steadily increased [5].

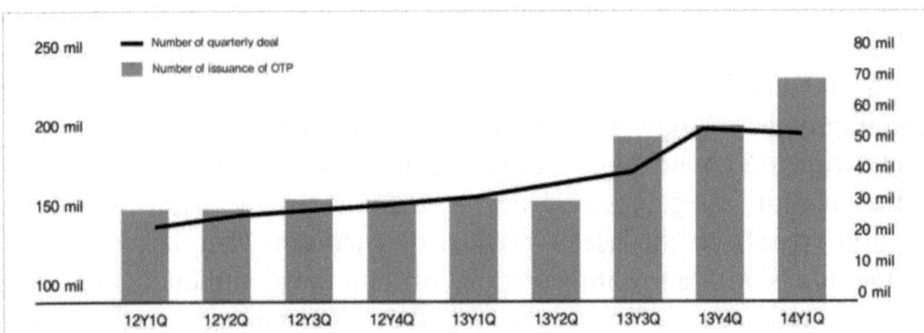

Fig. 1. Number of quarterly deal and number of issuance of OTP

The OTP authentication is the authentication means used to generate a one-time password by using the user's OTP token. Therefore, users have no uncomfortable to remember the password and the user's information is protected against the cyber threat generated in the process of password authentication.

Table 1. Comparison of the OTP algorithm

Algorithm	**HOTP** HMAC-based One-Time Password	**TOTP** Time-based One-Time Password	**OCRA** OATH Challenge-Response Algorithm
Synchronization Methods	**Event Synchronization**	**Timing Synchronization**	**Question and Answer Synchronization**
IETF Standards	RFC 4226 (2005.12.)	RFC 6238 (2011.05.)	RFC 6287 (2011.06.)
Security	**Low** (Before using OTP-Valid, After losing OTP - Danger, Required Changing PIN)	**High** (Whether OTP use or not user - Valid time(until 60s, 30s))	**High** (Always different Challenge, The challenge is leaked - Don't care : safe)
Convenience	**High** (Whenever clicking button Generating OTP, Repetitive using is Comfortable)	**Normal/Middle** (In OTP valid time - Generating same, Repetitive using is banned)	**Low, discomfortable** (Must enter the challenge value to token)
Authentication function	User Authentication	User Authentication	User /Transaction Authentication

If the user uses OTP methods, user will have no uncomfortable to remember the password and there is a secure benefits that user's information is protected against a number of cyber security threats that may arise in the password authentication process.

Recently, development of Transaction OTP which is one of the Transaction Signing technology has brought a movement to alternative e-trading scheme through the existing certificate [6]. However, OTP which have a number of advantages in security must be obtained the OTP token by the users and have a costs problem and management problem, inconvenience problem that user always have to own the OTP token. Also, especially in the case of transaction OTP, there is a problem that the user must enter the transaction information to the small size of OTP token [7].

3 The View of Biometric Technology

Recently, the Biometric authentication is attracting attention as authentication system on the next generation instead of knowledge-based authentication and possession-based authentication. And in the reverse side of the Biometric authentication, there is a 'smart revolution'. In case of World, the end of 2009, approximately 7 billion people who subscribed the number of smartphone has been exploded to approximately 16billion by 2014 [8].

And recent years, sensors as a fingerprint scanner that can collect the biometric information are included in the smartphone and it is resolving the safety problem from consignment of biometric information and issue of sensor's cost interfering with the activation of biometric-based authentication.

In fact, a number of smartphone such as the Pantech Vega LTE-A, the Apple iPhone 5S, Samsung's Galaxy S5 etc. are commercializing. And when the user uses the mobile-checking, the user can authenticate personally using the fingerprint authentication in recently years [9]. On this wise, Biometric authentication is beginning to activate by meeting with the smartphone and biometric authentication that have an advantage of convenience and safety will be more and more activated than the Knowledge-based authentication and Possession-based authentication.

Table 2. The comparison of smartphone's fingerprint authentication

	VEGA LTE-A	**iPhoine 5S**	**Galaxy S5**
Sensor Position	Back side button	Home button	Home button
Authentication methods	Up→down rubbing	Touch in all direction	Up→Down rubbing
	Registration of two fingerprint	Registration of five fingerprint	Registration of three fingerprint
Mobile checking	unsupported	AppStore, iBooks	Paypal
Fingerprint sensor	Crucialtec	Apple (Authentic takeover, 2012)	Synaptics

4 FIDO Authentication Protocol

FIDO(Fast Identify Online) Alliance is an association established to determine the technical standards for authentication using biometric technology in online environment. The alliance is founded in July 2012 and Paypal, Lenovo, Nok Nok Labs, Validity Sensors, Infineon, Agnitio are founding members [10]. By the November 2014, this alliance is comprised of 153 vendor (Google, MicroSoft, Samsung, Qualcomm, RSA, VISA and so on), including 21 Board members.

As described in Chapter 1 and Chapter 2, in the case of existing online authentication, mostly it is based ID/Password methods. So ID/Password is vulnerable to security, but there is not a perfect solution. Certificate also has a vulnerability in view that they still use the Password and the inconvenience which user must always hold the certificate exist.

On the other hand, Smart phone such as smart devices is not easy to enter the password because of the small screen. For this reason, the trend in smartphone was to enter a PIN number that performs user authentication or to draw a pattern. In recent years, biometric authentication techniques have been applied such as voice recognition and fingerprint recognition, so the user can authenticate simply and securely.

Using the biometric technology in smart devices, FIDO Alliance aims to provide simple and strong authentication methods for fast user authentication in online. That is, in a particular web server without using a password to authenticate a user, it is the way the device can authenticate the user on behalf of the web server.

FIDO standard proposes the following two protocols. The first is a technique for authenticating a user through UAF(Universal Authentication Framework) protocol interworking the authentication method applied from user's device with online service. The second is the U2F(Universal 2nd Factor) protocol that is able to add the second strong authentication in online authentication process, when the user logins the system [11].

4.1 FIDO Components

As shown as figure 2, FIDO is configured as a protocol to be transmitted between user device and server. FIDO client that is included in the user device performs the authentication on behalf of a majority of client programs. FIDO client and various authenticator are interworked in the abstract step of authentication token, token authentication API. That is, through the API, any kinds of authenticator can be compatible with FIDO Client.

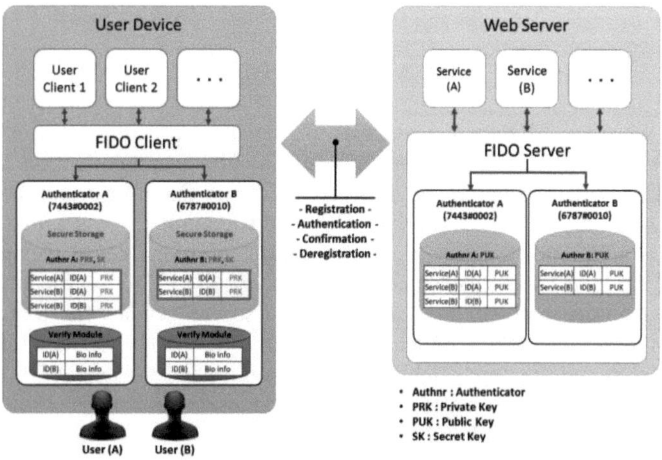

Fig. 2. FIDO components

Each authenticator have an identifier for each model and there exists a public key/private key pair corresponding there to. Among them, the private key is stored as an encrypted secret key inside the authenticator and the public key is stored in the FIDO server. Further, biometric information to verifying the user is encrypted and there are public key, private key pair corresponding to a service and user. Among them, the private key is stored as an encrypted secret key inside the authenticator and the public key is stored in the FIDO server.

4.2 Registration Protocol

In FIDO protocol, the process of registering the service ID and the public key corresponding to the user ID in the FIDO server is shown in Fig 3.

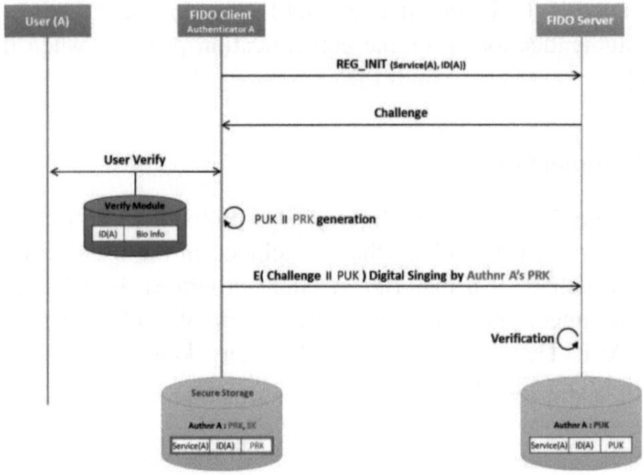

Fig. 3. Registration Protocol

1) First, when the FIDO client requests the registration using a service ID and user ID to the FIDO server, 2) FIDO server sends an available authentication list according to a randomly challenge value and security policy. 3) The user select the one from a list of authenticator, and the user go through a certification process with authenticator. 4) If the user authentication is successful, public key/private key pair is generated. 5) The public key and Challenge value received from the FIDO server is signed as an electronic signature with private key, 6) FIDO server verifies the digital signature using a public key of the authenticator. 7) Finally, the user's private key is stored in the FIDO Client and the user's public key is stored in the FIDO server.

4.3 Authentication and Confirmation Protocol

Figure 4 is the certificate process verifying the digital signature by the private key corresponding to the service ID and user ID in FIDO protocol.

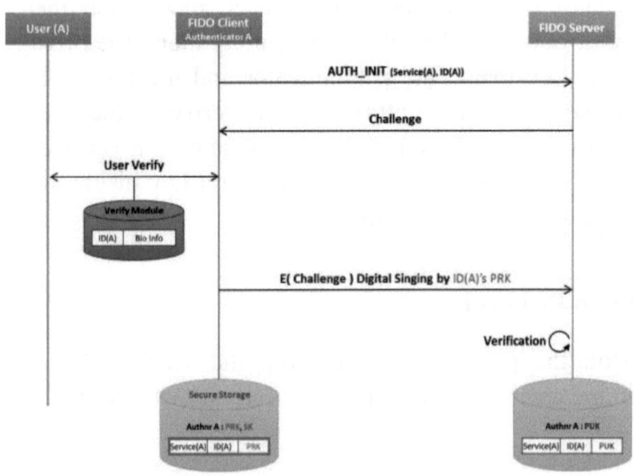

Fig. 4. Authentication & Confirmation Protocol

1) First, FIDO Client sends the authentication request to the service ID and user ID to the FIDO server, 2) FIDO server sends a random challenge value to FIDO client. 3) FIDO client display the authenticator list to the user, the user go though the certification process from the authenticator. 4) If the identity authentication was successful, FIDO client will unlock the private key corresponding to the service ID and user ID and the client will digital sign the Challenge value. 5) FIDO server identifies the user by verifying the digital signature value after finding the public key.

5 Conclusions

ID/Password that one of the knowledge-based authentication method is currently the most commonly used authentication method that have advantages of low cost and easy to implement and so on. But a number of social engineering problem can be happened in the trade-off relationship between the use and safety, it is vulnerable to a variety of cyber security threats. Although OTP have many advantages in security, there are cost problems that require the user to purchase and to manage a user OTP token, and inconvenience that the user must have the OTP and so on. Recently, with the diffusion of smart device equipped the knowledge-based sensor of the biometric recognition, biometric-authentication are attracting attention for reason of that the biometric authentication will replace information-based authentication and possession based authentication.

Like so, FIDO standard using the biometric authentication techniques is providing simple and strong authentication to fast user authentication in online. The advantage of FIDO is as follows. First, in the server side, because the user store the user's public key only, it is safe from the leak in the server Password hacking accidents. Second, in the network side, Challenge value is changed in every authentication process through the Challenge-Response methods, reusing the digital signature is impossible despite of the sniffing. Thirdly, in the client side, on the assumptions that authenticator is safely developed, the biometric information is safe from privacy issues because the user's biometric information is not exposed to the outside Authenticator.

However, as the user's public key corresponding to the "Authenticator ID-service ID-user ID" is stored in the server FIDO, performance issues may arise because the amount of storage that is the public key so much. In addition, since the private key corresponding to the "Service ID-User-ID" is stored in the authenticator, the convenience problems that the user must re-register all the service can be happened in the event of loss / failure of the authenticator. This disadvantage is not suitable for application in the authentication protocol FIDO wearable computing environment.

Ferguson Hill top prize in charge of financial payment services company PayPal (CPO) expect that the core content of wearable devices will be the mobile payment rather than health care and fitness. Developed countries, including the Unite States are expected that mobile payments will be fully replace a plastic credit cards in 2020. And Market research firm Gartner said in a recent report, the US mobile payment market sales were expected to grow from 23.5 million last year to $ 7,200 billion in

2017 [12]. However, in order to activation in/of mobile payment service market, it is expected that studies on the wearable authentication protocol for secure and convenient computing should be continued.

Acknowledgments. This research was partly supported by the R&D program of MSIP (Ministry of Science, ICT and Future Planning)[Project no. 10047528] and has been supported by the National GNSS Research Center program of Defense Acquisition Program Administration and Agency for Defense Development.

References

1. Lamport, L.: Password authentication with insecure communication. Communication of the ACM 24(11) (1981)
2. Shaw, R.: Social Engineering: A Hacking Story? Posted in Hacking on September 23, 2013. InfoSec Institute (2013)
3. Childress, R.L., Goldberg, I., Lechtman, M., Medini, Y.: Patent Issued for User Policy Manageable Strength-Based Password Aging. Journal of Engineering (February 20, 2013)
4. Owens, J., Matthews, J.: A Study of Passwords and Methods Used in Brute-Force SSH attack. Citeseer (2008)
5. Raddum, H., Nestås, L.H., Hole, K.J.: Security Analysis of Mobile Phones Used as OTP Generators. In: Samarati, P., Tunstall, M., Posegga, J., Markantonakis, K., Sauveron, D. (eds.) WISTP 2010. LNCS, vol. 6033, pp. 324–331. Springer, Heidelberg (2010)
6. Tsai, C.-L., Chen, C.-J., Zhuang, D.-J.: Trusted M-banking Verification Scheme based on a combination of OTP and Biometrics. Journal of Convergence 3 (September 2012)
7. Lin, M.H., Chang, C.C.,, A.: secure one-time password authentication scheme with low-computation for mobile communications. ACM SIGOPS Operating Systems Review (2004)
8. Emarketer: Smartphone Users Worldwide Will Total 1.75 Billion in 2014, January 16 (2014)
9. Tiwari, A.K., Agarwal, R., Goyal, S.: Biometric Authentication for Mobile Banking Security. Available at SSRN 2438213 (2014)
10. Fido Alliancess, `http://en.wikipedia.org/wiki/FIDO_Alliance`
11. Fido Alliance – Specification, `https://fidoalliance.org/specifications/overview/`
12. Gartner. Global mobile payment transaction volume from 2010 to 2017

SLAN Based User-Customized Cloud Interface Sharing for Smart Mobile Devices

Sanghyun Park, Jisu Kim, and Jinsul Kim*

School of Electronics & Computer Engineering, Jeonnam National University, Gwangju, 500-757, Korea
{sanghyun079,dyrk10}@gmail.com, jsworld@jnu.ac.kr

Abstract. Recently, the keyboard interfaces variously have been provided such as the hardware interface and the software interface. With smart mobile devices is supported with some software interfaces, but the number of interfaces are limited. This paper proposes an idea to provide dynamic interface for keyboard interface of smart mobile devices. The users can create their own interfaces or use the interfaces which are provided by server. We also propose SLAN communication, that help the interfaces can be easily shared between mobile devices. The user's IP or unique device name without knowing 1:1 or 1: N communication can be shared. Using SLAN to transmit the interface between smart mobile devices more quickly, efficiently than telecommunication that was used previously, and this is demonstrated through experiments.

Keywords: Soft keyboard, Mobile interface, SLAN, Dynamic interface.

1 Introduction

Smart mobile devices are used in the keyboard interface on a PC and a hardware interface, in the QWERTY format, users are more comfortable, and their taste in order to change in accordingly with the keyboard interface, utilizes a variety of applications. However, the application that provides the interface for the background and the keyboard can be changed variously, but the layout of the keyboard by the user themselves cannot easily change. In addition, the use of smart mobile devices, people use a variety of keyboard interface which is too inconvenient. We are using smart mobile devices, many people who provide the appropriate interface dynamically. Further, we provide the interfaces to offer by SLAN communication, SLAN using the wide range of people so that the interface can be shared. We have the comparison test performance of SLAM with existing smart mobile communication such as Bluetooth, DLAN [1-3].

* Corresponding author.

K.J. Kim and N. Wattanapongsakorn (eds.), *Mobile and Wireless Technology 2015*,
Lecture Notes in Electrical Engineering 310, DOI: 10.1007/978-3-662-47669-7_24

2 Related Work

There are many applications to provide a dynamic interface, but don't has papers and related materials are. We have a variety of cases, Kii Keyboard application description. Figure 1 shows the application of Kii Keyboard that has features a user-defined interface will be able to create. The user can set the desired label and color, an interface for storing the set, at any time, may be used. We proposed SLAN that users can create their own interfaces and share the interface can be easily with each other.

Fig. 1. Kii Keyboard Application

3 Dynamic Soft-Interface in Cloud Data Center

We provide a variety of interfaces through a server interface provides a production tool. As shown in Figure 2, using HTML5 [4-6] user through a web browser, the interface can easily be produced. Figure 2 shows the left side of the figure is an illustration of a user interface, making the interface by using a mouse or a finger can be adjusted to the size, the position can be adjusted. Additionally, the interface can be set to the value. After making the interface for the user to be stored in cloud data center as shown in Figure 2 the interface will come out of the list. To the right of the user interface for selecting and storing is that if smart mobile devices may be provided through the selected interface.

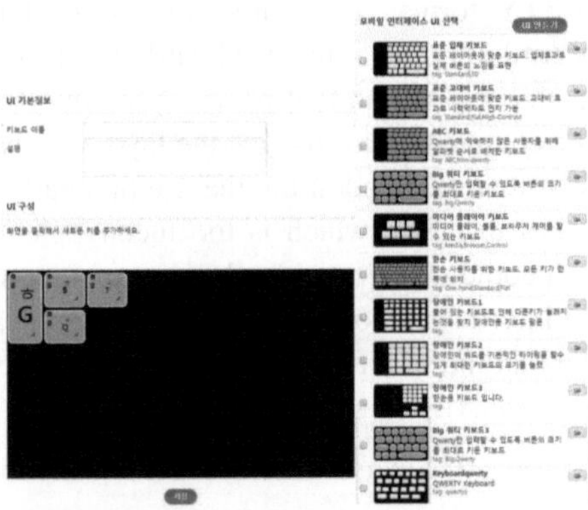

Fig. 2. Interface creation tools

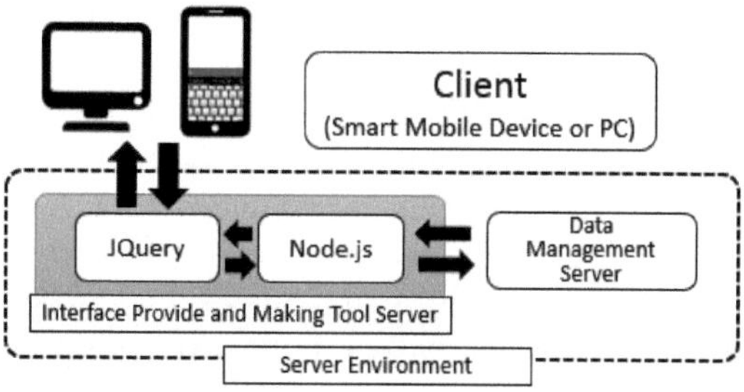

Fig. 3. Server Environment

Figure 3 shows the interface that provides the server with Node.js [7] web server function, and helps create the user interface feature that is JQuery [8]. When the interface is stored in the data management server through Node.js being stored, the user has to have an interface like the one shown in Figure 3 in order to fetch more data. In order to provide a wide range of application to dynamically use the following equations:

$$K_{width} \cdot \frac{D_{width}}{S_{width}} = R_{width} \tag{1}$$

$$K_{high} \cdot \frac{D_{high}}{S_{high}} = R_{high} \tag{2}$$

Smart mobile device varied as the internal shape of the resolution of each device different. Equation (1) and (2) are provided for the interface to match the resolution of calculation made in the keyboard K is the size of the production tool, D is the resolution of the mobile device user value, S refers to the resolution of a Web authoring tool, R is the mobile device user says the key size is applied. By calculating the high and the width of smart mobile device screen, the interface is provided.

4 Sharing Interfaces through SLAN

We have proposed SLAN for offer user interface can be shared easily compare with existing smart mobile Bluetooth, DLAN have disadvantages of the security.

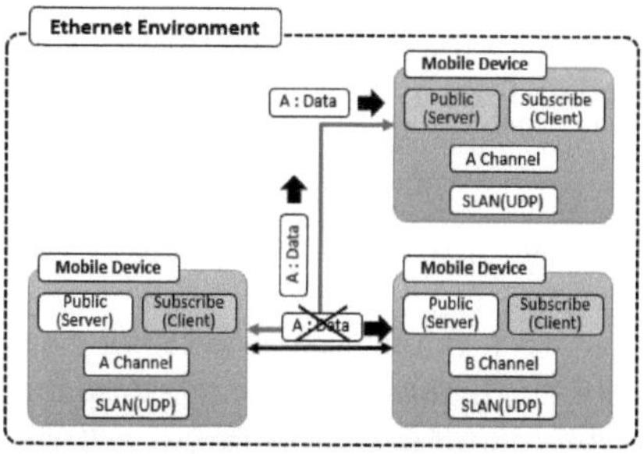

Fig. 4. SLAN Communication Structure

As shown in Figure 4, SLAN transmits the data in the ethernet environment with mobile devices, the Server and Client functions. The user's IP or unique device name, as shown in the figure don't know each other through the channel connection. When creating the first channel smart mobile devices that the server is later with the same channel to other mobile devices, when connected to the client. A user-interface for the transmission of the channel, the same channel interface to receive the mobile device, but other channels of the mobile device is not provided interface. SLAN is based on Wi-Fi transmit data so the transmission rate is fast. We prove that the performance SLAN is better than DLAN, Bluetooth communication.

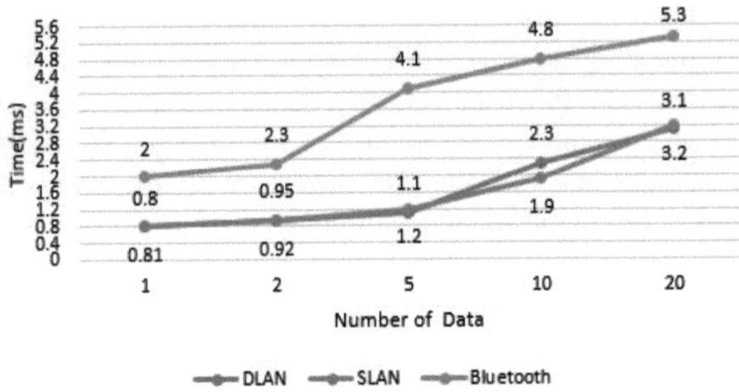

Fig. 5. Testing of Communication Speed

Figure 5 is a testing of DLAN, SLAN Bluetooth, respectively, and the interface data 1, 2, 5, 10, 20 sending data to ten average results for calculating a graph. Shows the result of the slower of Bluetooth and DLAN than SLAN and Wi-Fi is used because the speed difference. However, SLAN is the other party's IP or can transmit data without knowing the device name, different from 1: 1, as well as the DLAN 1: N multiple people have the advantage of being able to transfer data. Also we use DLAN the shortcomings of the speed data that can be transferred demonstrated.

5 Conclusions

We use a data center in cloud server where storage interfaces is created by users, the user can use smart mobile devices to provide dynamically interface for other users. Suitable user interfaces for their smart mobile are provided more efficiently. In addition, user can quickly share the various interfaces using the SLAN. The SLAN can transfer in faster speed and this is proved through experimentations. Users might not know each other's IP or device name was able to transmit data, and a variety of mobile devices could be applied dynamically through interface. Future work, the research is expected to continue with change in the images of interface shape.

Acknowledgements. This work (Grants No.C0218369) was supported by Business for Cooperative R&D between Industry, Academy, Research Institute funded Korea Small and Medium Business Administration in 2014-2015. Also, it was supported by Korea Evaluation Institute of Industrial Technology, Korea, 2014-2015.

References

1. Tamura, T., Masuda, I.: Device Connectivity Technologies Using Short-distance Wireless Communications. FUJITSU Sci. Tech. J., 213–219 (2013)
2. Shimizu, K., Suzuki, H., Watanabe, A.: Remote DLNA Communication System Based on NTMobile. In: 2012 IEEE 1st Global Conference on Consumer Electronics (GCCE), Tokyo, Japan, pp. 130–134 (2012)
3. Lo, S.-C., Yu, T.-H., Tseng, C.-C.: A Remote Control and Media-Sharing System Using Smart Devices. Journal of Systems Architecture, 1–13 (2014)
4. Cha, S.-H., Yun, Y.: Smartphone Application Development using HTML5-based Cross-Platform Framework. International Journal of Smart Home 7, 195–201 (2013)
5. Melamed, T., Clayton, B.: A comparative evaluation of HTML5 as a pervasive media platform. In: Phan, T., Montanari, R., Zerfos, P. (eds.) MobiCASE 2009. LNICST, vol. 35, pp. 307–325. Springer, Heidelberg (2010)
6. Padley, R.: HTML5-bridging the mobile platform gap: mobile technologies in scholarly communication. The Journal for the Serials Community 24, 32–39 (2011)
7. Charland, A., Leroux, B.: Mobile application development: web vs. native. Communications of the ACM 54, 49–53 (2011)
8. Zibula, A., Majchrzak, T.A.: Cross-Platform Development Using HTML5, jQuery Mobile, and PhoneGap: Realizing a Smart Meter Application. In: Cordeiro, J., Krempels, K.-H. (eds.) WEBIST 2012. LNBIP, vol. 140, pp. 16–33. Springer, Heidelberg (2013)

Author Index

MIX
Papier aus verantwortungsvollen Quellen
Paper from responsible sources
FSC® C105338

If you have any concerns about our products,
you can contact us on
ProductSafety@springernature.com

In case Publisher is established outside the EU,
the EU authorized representative is:
Springer Nature Customer Service Center GmbH
Europaplatz 3, 69115 Heidelberg, Germany

Printed by Libri Plureos GmbH
in Hamburg, Germany